Valuing Machinery and Equipment: The Fundamentals of Appraising Machinery and Technical Assets

by
Machinery and Technical Specialties Committee
of the American Society of Appraisers

The International Society of Professional Valuers

Library of Congress Cataloging-in-Publication Data

American Society of Appraisers.
Valuing machinery and equipment / by the American Society of Appraisers.
p. cm.
Includes index.
ISBN 0-937828-04-1
1. Industrial equipment—Valuation—United States. I. Title.
HD39.35 .A535 2000
657'.73—dc21
00-009888

ISBN 0-937828-04-1

Printed in the United States of America.

For information about the American Society of Appraisers, call (800) 272-8258 or (703) 478-2228, or write American Society of Appraisers, P.O. Box 17265, Washington, DC 20041-0265.

Table of Contents

Preface

The occupation of equipment valuation precedes the beginning of the twentieth century, yet very little has ever been written regarding its theory, concepts, and practice.

Nearly all appraisal work prepared in the first half of this century was insurance related. Appraisers were typically trained by appraisal or insurance companies utilizing apprenticeship programs that often lasted up to five years. There was little uniformity in terms of procedure between companies, as standardized appraisal techniques for equipment had not been developed. Each senior appraiser had his own methods and ideas as to how equipment should be valued, and was often reluctant to pass "hands on" knowledge to those wishing to enter the field. For the sharing of knowledge and information, it was a dark time for equipment appraisers.

It was not until 1936, when the Iowa State University Press published what has become a classic: *Engineering Valuation and Depreciation,* by Marston, Winfrey, and Hempstead, that some headway was made. In the Preface of their book, the authors stated "the complex society of the present demands systematic and theoretically correct procedures when consideration is given to the appraisal of enterprises and properties not regularly acquired on the market."

This is truer than ever in today's global economy. Technologies exist today that were undreamed of in 1936. With the application of proper procedures, guidelines and appraisal theory developed in a large part by the American Society of Appraisers and the Appraisal Foundation, the most complex and esoteric of industries can be valued by the properly trained appraiser.

In 1968, *Appraisal Principles and Procedures*, by Dr. Henry A. Babcock, FASA, was published by the American Society of Appraisers. This text, and *Engineering Valuation and Depreciation,* became the recognized sources of valuation theory for the equipment appraiser.

A Machinery & Equipment Committee (now known as the Machinery and Technical Specialties Committee) was formed within the American Society of Appraisers in 1983. One of its primary goals was to publish a hardcover textbook that would specifically address theories and concepts of equipment valuation not covered in the earlier books. This text, *Appraising Machinery and Equipment,* was published by the Society in 1989.

As part of its educational efforts, the Committee, in conjunction with the Society, went on to develop a series of Level Courses to teach the theory and practice of equipment valuation. The Committee decided in 1994 that a new textbook would be necessary to bridge the gap between the earlier work and the Level Courses that addressed more advanced subjects. It is the belief of the Committee that this text will meet the requirements of the Machinery and Technical Specialties discipline for the immediate future.

With the new text, the Level Courses and the many fine articles published in the Committee's Newsletter, *The MTS Journal,* introduced in 1984, the light of knowledge is finally penetrating what was for too many years a dark and hidden field of endeavor.

The astronomer, Percival Lowell, was quoted as having said: "It takes about twenty years for new knowledge to filter into the textbooks." Unfortunately, in the field of equipment appraising it has taken far longer. In publishing this text and the earlier 1989 book, the American Society of Appraisers is doing its best to rectify this perception.

H. Denis Neumann, ASA
Emeritus Member
Machinery & Technical
Specialties Committee

Acknowledgments

The writing and preparation of this book is the culmination of the efforts of a number of appraisal professionals. These individuals have contributed countless hours to this project and it is their dedication and commitment to the profession that have made this manuscript possible. From writing to reviewing, advising to critiquing, each one of the individuals mentioned below has provided valuable input in one form or another.

Special thanks are reserved for

Robert B. Podwalny, ASA, who in 1994 as the Chairman of the then Machinery and Equipment Committee of the American Society of Appraisers, initiated the task of writing the book;

Gerald L. Huether, ASA, who is the current Chairman of the Machinery and Technical Specialties Committee and served as the technical editor;

Kent L. Osborne, ASA, who reviewed and edited the entire text; and

Rebecca Maxey, Director of Publications, American Society of Appraisers, who coordinated and oversaw the publication process.

Again, I wish to thank the following individuals for their significant contributions to the society and the profession:

Leslie H. Miles, Jr., ASA
Michael J. Remsha, ASA
Robert S. Svoboda
Jackie L. Montalvo, ASA
Jeffrey A. Hutton, ASA
J. Michael Clarkson, ASA
John J. Connolly, III, ASA
Melvin I. Fineberg, ASA
Eugene G. Kaczkowski, ASA
David N. Lang, ASA
Doran V. McClellan, ASA
Doby A. Rose, ASA
Alan C. Iannacito, ASA
John Madge, FASA
H. Denis Neumann, ASA
Paul Rice, ASA
Ellis French
Susan Brady
Jeff Brady
Raymond A. Springer, ASA
Dennis C. Neilson, ASA

Also, a grateful acknowledgment is due to the many ASA reviewers who contributed their valuable comments during the review process.

Arthur E. Narverud, Chairman
MTS Textbook Subcommittee

Disclaimer

The book is not intended as an advanced text or as a definitive or complete statement of the theory and methods of appraising machinery, equipment, and industrial assets and facilities. Such a definitive statement cannot be written at this time. Inclusion of certain methods and theories in the book does not mean, and is not to be interpreted as meaning, that other methods and theories not mentioned are disapproved. It follows that the individuals named as having worked on the book or on the peer review committee do not agree with everything in the book. In fact, disagreements between those writing and editing the book were common. It could not be otherwise since the theory and methods of appraising machinery, equipment, and industrial assets and facilities is a developing field.

1

Introduction

Objectives:

1. Define important terms.
2. State the basic concepts of value.
3. Define the different premises of value used by machinery and technical specialties (MTS) designated appraisers.
4. Introduce the three approaches to value.
5. Differentiate price, cost, and value.
6. Contrast accounting and valuation depreciation.
7. State appraisal purposes, objectives, and events that influence appraisal value.
8. Describe the types of clients encountered by MTS appraisers.

Definition of Appraisal and Value

An *appraisal* is the act or process of determining *value*.[1] *Value* has been defined as the monetary worth of property, goods, or services.[2] The American Society of Appraisers (ASA) has broadened the definition of appraisal to include any of the four following operations, independently or in combination:

1. Determination of the value of property.

2. Estimation of the cost of (a) production of a new property, (b) replacement of an existing property by purchase or production of an equivalent property, or (c) reproduction of an existing property by purchase or production of an identical property.

3. Determination of the nonmonetary benefits or characteristics that contribute to value. The rendering of judgments as to age, remaining life, condition, quality, or authenticity of physical property.

4. Forecast of the earning power of property.[3]

The appraisal of machinery, equipment, and certain other business assets[4] encompasses all four of these meanings.

In this book, the terms *appraisal* and *valuation* are used interchangeably. *The Uniform Standards of Professional Appraisal Practice* (USPAP) distinguish between *appraisal* and *consulting*. USPAP defines *consulting* as the act or process of providing information, analysis of real estate data, and recommendations or conclusions on diversified problems in real estate, other than an opinion of value.[5] For some purposes, the distinction between appraisal and consulting is important, but in this book, we will generally not be concerned with making fine distinctions between the two activities.

Premises of Value Relating to MTS Assets

Because the MTS appraiser deals with a variety of assets, most of which can be moved, it is necessary to recognize different premises of value. These can be broadly classified into three categories, distinguished mainly by an asset's anticipated use:

1. Sale for removal for a similar or alternate use.

2. Continued (or installed) use of the asset for the purpose for which it was designed and acquired.

3. Liquidation.

Hence, use of the term *value* or *fair market value* is modified or refined to create special definitions to fit the needs of a particular appraisal. These modifiers provide a specific premise of value to guide the work of the appraiser. The following list of the various definitions of *value* that the appraiser will encounter is not intended to be complete. The list begins with a definition of *fair market value* and then presents the various refinements of the term that are used to fit the needs of the appraiser.

Sale for Removal or Alternate Use

Fair market value is the estimated amount, expressed in terms of money, that may be reasonably expected for a property in an exchange between a willing buyer and a willing seller, with equity to both, neither under any compulsion to buy or sell, and both fully aware of all relevant facts, as of a specific date.[6]

Fair market value—removal is the estimated amount, expressed in terms of money, that may reasonably be expected for a property, in an exchange between a willing buyer and a willing seller, with equity to both, neither under any compulsion to buy or sell and both fully aware of all relevant facts, as of a specific date, considering the cost of removal of the property to another location.

Continued Use (or Capacity for Use)

Fair market value in continued use is the estimated amount, expressed in terms of money, that may reasonably be expected for a property in an exchange between a willing buyer and a willing seller, with equity to both, neither under any compulsion to buy or sell, and both fully aware of all relevant facts, including installation, as of a specific date and assuming that the business earnings support the value reported.[7] This amount includes all normal direct and indirect costs, such as installation and other assemblage costs to make the property fully operational.

Fair market value—installed is the estimated amount, expressed in terms of money, that may reasonably be expected for an installed property in an exchange between a willing buyer and a willing seller, with equity to both, neither under any compulsion to buy or sell, and both fully aware of all relevant facts, including installation, as of a specific date. This amount includes all normal direct and indirect costs, such as installation and other assemblage costs, necessary to make the property fully operational.

Liquidation

Orderly liquidation value is the estimated gross amount, expressed in terms of money, that could be typically realized from a liquidation sale, given a reasonable period of time to find a purchaser (or purchasers), with the seller being compelled to sell on an as-is, where-is basis, as of a specific date.

Forced liquidation value is the estimated gross amount, expressed in terms of money, that could typically be realized from a properly advertised and conducted public auction,[8] with the seller being compelled to sell with a sense of immediacy on an as-is, where-is basis, as of a specific date.

Liquidation value in place is the estimated gross amount, expressed in terms of money, that could typically be realized from a failed facility, assuming that the entire facility would be sold intact with a limited time to complete the sale, as of a specific date.

These definitions are not the only acceptable ones. Certain appraisals require the use of definitions mandated by law. Definitions may be expanded or redefined by the appraiser as the purpose and function of the appraisal dictate, so long as the fundamental and underlying concept is not altered without a compelling reason (such as being required by law).

Other important value premises or definitions include the following:

> *Salvage value* is the estimated amount expressed in terms of money that may be expected for the whole property or a component of the whole property that is retired from service for use elsewhere.
>
> *Scrap value* is the estimated amount expressed in terms of money that could be realized for the property if it were sold for its material content, not for a productive use.
>
> *Insurance replacement cost* is the replacement cost new as defined in the insurance policy less the replacement cost new of the items specifically excluded in the policy, if any.

Insurance value depreciated is the insurance replacement cost new less accrued depreciation considered for insurance purposes, as defined in the insurance policy or other agreements.

Price and Cost Distinguished

The appraiser often receives financial information about the property to be appraised in which the terms *price, cost,* and *value* are used interchangeably by nonappraisers. Appraisers carefully distinguish between these terms.

Price is defined as the amount a particular purchaser agrees to pay and a particular seller agrees to accept under the circumstances surrounding their transaction.[9] *Price* implies an exchange and is an accomplished fact.[10] The price paid for a particular asset may be higher or lower than, or equal to, the asset's value.

The term *cost* is used by appraisers in relation to production, not exchange. Cost may be either an accomplished fact or a current estimate.[11] It is defined as the total dollar expenditure for any asset.[12] The cost of a particular asset may be higher than, lower than, or equal to the asset's value.

Approaches to Value

There are three generally recognized approaches to the determination of value: cost, sales comparison, and income.[13] These approaches are widely accepted by financial institutions, courts, government agencies, business, and society in general, and they establish theoretical concepts and systematic methods. These approaches are briefly defined here and are discussed in detail in later chapters.

- **Cost approach:** The appraiser adjusts the replacement cost (new) of the asset being appraised for the loss in value caused by physical deterioration, functional obsolescence, and economic obsolescence. The cost approach is based on the principle of substitution: a prudent buyer will not pay more for an asset than the cost of acquiring a substitute property of equivalent utility.[14]

- **Sales comparison approach:** The appraiser adjusts the prices that have been paid for assets comparable to the asset being appraised, equating the comparables to the subject.
- **Income approach:** The appraiser determines the present value of the future economic benefits of owning the property.

Although USPAP requires that all three approaches to value be considered, the valuation of certain assets or the valuation premise under consideration may make the use of all three approaches impractical. For example, the cost and income approaches are generally of little use in determining a liquidation value. The cost approach, without sufficient research and quantification of depreciation and obsolescence, may not accurately reflect the fair market value of a particular asset. It may not be possible to use the income approach to determine the in-use value of equipment because it may be impossible to isolate the income attributable to the equipment. The sales comparison approach would be impossible to use for a one-of-a-kind machine that has never been exposed to, or sold in, the marketplace. There are many other examples. The valuation circumstances involving a particular asset may not allow the application and correlation of all three approaches to value. This is consistent with Standard Rule 7-4 of USPAP, which requires the appraiser to consider all three approaches to value and decide which approaches are applicable to the situation at hand.

Valuation Depreciation and Accounting Depreciation

Depreciation is another term that appraisers use differently from nonappraisers. In particular, the valuation concept of depreciation differs from the accounting concept of depreciation. Depreciation for valuation purposes is the estimated loss in value of an asset, compared with a new asset; appraisal depreciation measures value inferiority[15] that is caused by a combination of physical deterioration, functional obsolescence, and economic (or external) obsolescence.

It is important for the appraiser to understand that the accounting depreciation process is one of cost allocation only. It is not a method of valuation. Because a company's fixed assets are not held for resale, there is no attempt to reflect any change in the market value of the assets. As depreciation is calculated from period to period, it is added to an accumulated depreciation account. Depreciation for accounting purposes may be thought of as a mathematical procedure for recovering the original cost of an asset in consistent installments over a specified period.

Thus, the primary difference between the valuation and accounting concepts of depreciation is that appraisal depreciation measures value inferiority, whereas accounting depreciation is a mathematical convention for recovering an asset's cost. Valuation depreciation is covered in chapter 3, which discusses the cost approach.

Other Accounting Terms

The appraiser will often work with information furnished by the accounting or financial departments of manufacturing and commercial enterprises. The appraiser should have a basic understanding of accounting terms and their relation to valuation concepts.

In addition to *cost, price,* and *depreciation,* the appraiser encounters other accounting and financial terms. Some of these terms refer to the methods used to account for a company's assets. In the accounting process, the accountant either *capitalizes* or *expenses* the costs incurred or prices paid for certain items. If the expenditure is capitalized, the item is considered a *fixed asset,* and the appraiser finds the *original cost* of the asset (in the hands of its present owner) entered into the owner's fixed asset record. Its cost will be recovered through accounting depreciation. These procedures are designed to record costs in accordance with generally accepted accounting principles (GAAP). Tangible assets used in the operation of the business that are not intended for resale to its customers are generally referred to as plant, property, and equipment, or fixed assets.

Usually, the accounting record assigns the fixed assets to accounts for land, land improvements, buildings, and equipment. Except for land, these assets are depreciated over a predetermined service life. Land is retained at cost and is not depreciated on the accounting record.

Again, it is important to understand that the accounting depreciation process is one of cost allocation only; it is not a method of valuation. *Accumulated depreciation* represents the total accounting depreciation taken against the cost of an asset as of a given date and depends on the accounting method used. Subtracting accumulated depreciation from cost yields the *net book value,* which is the capitalized cost of an asset less the accounting depreciation taken for financial reporting. It is derived through a cost allocation process, not a valuation process.

The cost of the asset may or may not be the same as its *depreciable cost,* which is the accounting cost of a property less the estimated salvage value that can be obtained for the property at the end of its useful life. The *salvage value* is the estimated amount, expressed in terms of money, that may be expected on sale or other disposition of an asset that is retired from service after it is no longer useful. Again, it is important to note that in using the term *value,* selected modifiers clearly define the *type* of value. On rare occasions, the net book value may approximate an appraisal value; however, that occurs only by chance.

The *life* of the asset on a company's books is also an accounting concept. In theory, the depreciable or book life for GAAP purposes should reflect the company's estimate of the asset's expected useful life, whereas the tax life is prescribed by tax regulations and may differ materially from the asset's expected useful life. However, the appraiser will find companies that choose to base their GAAP (i.e., book) lives on the federal income tax cost recovery schedules, such as the Modified Accelerated Cost Recovery System. Under this system, companies are generally required to use lives of five to seven years for equipment (depending on the taxpayer's industry classification), 15 years for land improvements, and 39 years for buildings.

The amount of accounting depreciation taken over the life of an asset cannot exceed its cost. Once the net book value reaches zero, no more depreciation is taken. However, the "fully depreciated" asset may remain in service and still have a value for valuation purposes.

The Valuation Process

The valuation process is a systematic procedure used by appraisers to provide answers to their clients' questions about value and value-related issues. It begins when the appraiser fully understands and identifies the appraisal problem at hand and concludes when the appraiser reports his or her solution to the client. The number of steps taken to solve the problem depends on the nature of the appraisal engagement and the availability of data. In any appraisal assignment, the goal of the valuation process is to produce a well-supported opinion of value showing that the appraiser has considered all factors that materially affect the value of the assets being appraised.

The valuation process can be summarized as follows: the appraiser (1) assembles relevant data; (2) conducts market research; (3) applies appropriate analytical techniques, knowledge, experience, and judgment to reach conclusions; (4) formulates value or other conclusions; and (5) prepares an appraisal report of some type. The following outline shows the steps in the valuation process in greater detail:

1. Define the problem.
   - 1-1. Property to be appraised
   - 1-2. Purpose of appraisal
   - 1-3. Intended uses of appraisal
   - 1-4. Premise of value
   - 1-5. Effective date
   - 1-6. Limiting conditions
2. Collect relevant data.
   - 2-1. List, describe, and classify assets.
   - 2-2. Research cost, comparable sales, and income data.

3. Apply appropriate valuation methodology and analytical techniques.
4. Formulate value (or other) conclusions.
5. Prepare appraisal report.

1. Define the Problem

Property to Be Appraised

What is the property to be appraised?[16] Also, what assets are excluded from the appraisal? It is important to identify both included and excluded assets, because the scope of the appraisal may not be obvious. Take for example a complicated appraisal subject such as a complex industrial plant. Is land excluded? If land is included, who will appraise it? Does the appraisal include assets at more than one location?

Purpose of Appraisal

The purpose of the appraisal is essential to the identification of any intended users of the report. Is the purpose to determine fair market value in continued use, orderly liquidation value, or some other kind or level of value? It is essential that any intended user of the appraisal understand the purpose.

Intended Uses of Appraisal

The intended use of an appraisal is established by the client. The client's use may encompass requirements of one or more other intended users. An appraiser cannot identify the client's intended use without having identified the client and having established a clear understanding of the client's requirements by communicating with the client or the client's agent.[17] To avoid misunderstandings, the appraiser and client must have a mutual understanding of the uses of the appraisal report before the appraiser begins to work. Some of the possible uses of an appraisal are discussed later in this chapter.

Premise of Value

The different premises of value needed by the MTS appraiser have been discussed above. Whereas the intended use of the appraisal is determined by the client, the value premise is determined by the appraiser, after clear communication with the client regarding the intended use and other circumstances surrounding the appraisal requirement.

Effective Date

The appraisal must be "as of" a specific date because values are constantly changing. The effective date is the date that correlates to the value conclusions, not the date when the appraisal was started or completed (although the effective date could be either of those dates). Sometimes a retroactive date is important, such as in inheritance tax cases (i.e., date of death).

Limiting Conditions

A report's "statement of limiting conditions" limits the use of an appraisal or qualifies its conclusions. In addition to clarifying the meaning of the report for intended users, the limiting conditions could be important in limiting the appraiser's liability.

2. Collect Relevant Data

List, Describe, and Classify Assets

The assets should be properly listed in a uniform and systematic manner, not only to facilitate pricing but also to assist the client or other intended user in identifying assets. The catchphrase "name it, size it, describe it" has been used to describe the basics of listing. Using the client's asset accounts and classes is often desirable, especially when the appraiser's work must be reconciled with the client's accounting records. This subject is discussed in chapter 2.

Research Cost, Comparable Sales, and Income Data

This subject is discussed in detail in chapters 3 through 5.

3. Apply Appropriate Valuation Methodology and Analytical Techniques

USPAP requires that the appraiser should *consider* all three approaches: the cost, comparable sales, and income approaches. It is not required that all three approaches be used, but the appraiser must consider each approach and state why those not used were omitted. The three approaches are the subjects of chapters 3 through 5.

4. Formulate Value (or Other) Conclusions

The appraiser's conclusion of value (or other conclusions if the assignment is a consulting engagement) must be clearly stated and the basis for the conclusion sufficiently explained.

5. Prepare Appraisal Report

The appraisal report should be detailed enough to convey the "what, where, when, and how" of the appraisal, among other things. The facts, reasoning, and methodology used to arrive at conclusions of value should be well organized, clear, and concise and should meet USPAP and ASA requirements. These subjects are covered in the discussions of report writing (chapter 6) and ethics (chapter 7).

Appraisal Purposes and Premise of Value

Appraisals of machinery, equipment, and other business assets have a wide field of application. Appraisals are required for a variety of purposes, including allocation of purchase price; bankruptcy; condemnation; dissolutions of marriage, partnerships, and corporations; financing; insurance; leasing; management considerations; mergers and acquisitions; partnership formation and dissolution; transfer of ownership; various types of taxation and tax planning; and utility rate making. Each appraisal purpose requires that an appropriate level, type, or premise of value be selected. As discussed earlier, the client defines the intended use of the appraisal, but it is the appraiser's responsibility to select the proper premise of value associated with the intended use of the appraisal.

The discussion that follows is general in nature and is not intended to include all the uses for which appraisals are prepared. The experienced appraiser has probably seen virtually all premises of value appropriately applied for virtually all purposes: for example, allocation of purchase price based on liquidation, lease residuals based on liquidation, or financing at market value. There are exceptions to the general rule; thus, the following discussion is intended to provide general guidelines only. Whenever possible, the appraiser should seek guidance from the client's legal counsel or accountant with respect to the appropriate premise or level of value.

Allocation of Purchase Price

Often, entities containing multiple assets are purchased as a unit. Subsequently, there may be a need to know the values of the separate components of the purchase in order to set up appropriate asset records, depreciation schedules, or similar information. Depending on the purpose of the allocation (e.g., federal income tax or financial accounting), the allocation may be based on "fair market value" under the Internal Revenue Code or "fair value" under Accounting Principles Board Opinions 16 and 17.[17]

Bankruptcy

A variety of value premises may need to be addressed in bankruptcy appraisals. In many cases, a liquidation premise is appropriate, because creditors usually prefer to be paid in cash, not equipment. A liquidation value appraisal will frequently estimate the likely net recovery from the forced sale of the assets. However, bankruptcy appraisals can also be required to address value-in-use, value-in-exchange, liquidation value, and net realizable value. The proper premise is a function of the facts and circumstances. An appraiser often needs to review prior court rulings or seek guidance of legal counsel to determine the appropriate premise or level of value.

Condemnation

When a public agency needs to acquire private property for conversion to public use, it exercises the power of eminent domain. Often, the property in question is industrial or commercial in character and may contain extensive machinery and equipment. The value premise or definition is mandated by the law of the jurisdiction in which the taking occurs. Generally it is a variant of fair market value, such as *fair market value in continued use,* but the appraiser should seek the guidance of legal counsel with respect to the appropriate premise or level of value.

Dissolutions of Corporations, Partnerships, and Marriages

Appraisals for the purpose of dissolutions, whether they are marriages, partnerships, or corporations, are often done on a fair market value basis to establish an equitable distribution to each spouse, partner, or stockholder. However, an appraisal for dissolution could possibly be done on a liquidation value basis, particularly if the involved parties wish to consider disposition of all assets first and then division of the monetary results. Again, the appraiser should seek the guidance of legal counsel with respect to the appropriate premise or level of value.

Financing

With respect to asset-based lending, when a lending institution extends credit to a customer, the collateral that is pledged by the borrower must be appraised. Strict rules usually govern the amount that may be borrowed and stipulate that the sum cannot exceed a specified amount of the appraised value. To protect the lending institution and to ensure a full recovery of the loan in the event of a default, the appraised value is often, though not necessarily, based on a liquidation value premise—that is, what could be recovered from the distress sale of the assets pledged to guarantee the loan. Lenders do not want the assets; they want cash, as soon as possible. In some cases in which the borrower has an excellent credit record and is well known in the local business community, a lending institution may allow the loan to be based on some type of market value premise.

Insurance and Loss Settlement

The determination of insurable value is a common purpose for which appraisals are required. Insurance appraisals are required to establish a value for insurance coverage to indemnify the insured against loss. The insurable value is of concern to owners, lessors, lessees, insurers, agents, and brokers.

The insurable value may be equivalent to replacement cost new if the insurance coverage is provided on a replacement policy. More often, however, the insurable value will be predicated on replacement cost new less depreciation or a used cost of like kind, quality, and condition plus appropriate freight, taxes, and installation. This is often referred to as the actual cash value.

Because many insurance policies covering industrial and commercial machinery and equipment include a coinsurance (average) clause, a good appraisal is essential to ensure financial recovery in case of loss.

An appraisal done for an insurance loss settlement has a very special and limited purpose: to verify that asset values are in compliance with insurance policy requirements. The values determined are the same as for an insurance appraisal. The only real difference is that for loss settlement the appraisal is done after the loss has occurred.

Leasing

The subject of leasing is discussed in chapter 8.

Management Considerations

This catchall appraisal purpose often relates to internal financial considerations. It is usually required by business owners. Management or owners may want to determine the economic worth of their business, possibly by using various hypothetical assumptions as a basis for analysis. The value premises would depend on the client's objectives.

Taxation—Income Tax

Income and other taxes are often the primary motivation for appraisals performed by the "Big 5" accounting firms and other large, independent appraisal companies. A complete discussion of the tax-related reasons for doing appraisals could easily take up a chapter and is beyond the scope of this book. There are many income tax-related reasons why a client would need an appraisal. One of the most common has already been mentioned: the buyer's (and the seller's) need to allocate the purchase price when complex industrial assets have been acquired. An issue that sometimes arises is whether "fair market value" under the tax code equates to *fair market value in continued use* or some other variant of fair market value. In these and other income-tax-related appraisals, it is essential for the lead appraiser to establish clear communication with the client's income tax personnel at the onset, thus ensuring that the correct value premise has been selected.

Taxation—Property Tax

Business property owners who think their property tax assessments may not be fair and reasonable often hire appraisers to render an opinion of value. Depending on the value conclusion reached, the appraisers may be retained to assist in negotiations with local assessment officials or to provide expert witness testimony if the dispute cannot be resolved through negotiation. Local law generally often mandates the premise of value that must be used. This will generally be a variant of fair market value, which often—but not always—equates to *fair market value in continued use* for the machinery and equipment.

Taxation—Estate and Gift Taxes, Charitable Contributions

Appraisals made for federal estate and gift tax purposes generally will be required by tax regulations to use "fair market value," the term used by the federal tax code. As always, the appraiser should seek guidance from the client's attorney or accountant. For example, as in income-tax-related appraisals, the issue sometimes arises as to whether "fair market value" under the tax code equates to *fair market value in continued use* or some other variant of fair market value.

Key Points

- An *appraisal* is the process of estimating or determining value. *Value* is the monetary worth of property, goods, or services.

- Because the MTS appraiser deals with a variety of assets, most of which can be moved, it is necessary to recognize different premises of value. These can be broadly classified into three categories, distinguished mainly by the asset's anticipated use: *sale for removal* for a similar or alternate use; *continued use* of the asset for the purpose for which it was designed and acquired; or *liquidation*. Hence, the MTS appraiser modifies or refines the words *value* or *fair market value* to create special definitions to fit the needs of a particular appraisal. These modifiers provide a specific premise of value to guide the appraiser in his or her work.
- The appraiser often receives financial information about the property to be appraised in which the terms *price, cost,* and *value* are used interchangeably by nonappraisers. Appraisers carefully distinguish between these terms. *Price* is the amount a particular purchaser agrees to pay and a particular seller agrees to accept under the circumstances surrounding their transaction. *Price* implies an exchange and is an accomplished fact. The price paid for a particular asset may be higher or lower than, or equal to, the asset's value. The term *cost* is used by appraisers in relation to production, not exchange. Cost may be either an accomplished fact or a current estimate. It is defined as the total dollar expenditure for any asset. The cost of a particular asset may be higher, lower than, or equal to, the asset's value.
- Three approaches to value are generally recognized: the cost, sales comparison, and income approaches. These approaches are widely accepted by financial institutions, courts, government agencies, business, and society in general. The valuation circumstances involving a particular asset may not allow the application and correlation of all three approaches to value. USPAP merely requires the appraiser to consider all three approaches to value and decide which approaches are applicable to the situation at hand.
- Depreciation is another term of art that appraisers use in a different way than nonappraisers. The valuation concept of depreciation is different from the accounting concept of depreciation. Depreciation for valuation purposes may be thought of as the estimated loss in *value* of an asset compared with a new asset. This value inferiority may be caused by a

combination of physical deterioration, functional obsolescence, and economic (or external) obsolescence. Accounting depreciation, on the other hand, is a process of *cost* allocation only. It is not a method of valuation.

- The valuation process is a systematic procedure used by appraisers to provide answers to their clients' questions about value and value-related issues. The goal of the valuation process is to produce a well-supported opinion of value that shows the appraiser has considered all factors that materially affect the value of the assets being appraised.
- Appraisals of machinery, equipment, and other business assets are required for a variety of purposes, including allocation of purchase price; bankruptcy; condemnation; dissolutions of marriage, partnerships, and corporations; financing; insurance; leasing; management considerations; mergers and acquisitions; partnership formation and dissolution; transfer of ownership; and various types of taxation and tax planning. Each appraisal purpose or intended use requires that an appropriate level, type, or premise of value be selected. The client defines the intended use of the appraisal, but it is the appraiser's responsibility to select the proper premise of value associated with the intended use of the appraisal. It is essential that the appraiser seek guidance from the client's attorney, accountant, or tax manager with respect to the appropriate premise of value *before* beginning the appraisal.

Additional Reading

The Appraisal of Real Estate. 11th ed. Chicago: Appraisal Institute, 1996.

Babcock, Henry A., FASA. *Appraisal Principles and Procedures.* Washington, D.C.: American Society of Appraisers, 1989.

Bonbright, James C. *The Valuation of Property.* New York: McGraw-Hill, 1965.

Notes

[1] *Uniform Standards of Professional Appraisal Practice* (USPAP), Definitions, 2000, p. 10. Appraisal Standards Board. The Appraisal Foundation, Washington, DC.

[2] The Dictionary of Real Estate Appraisal *(DREA),* 2nd ed., p. 318. American Institute of Real Estate Appraisers, Chicago, 1989.

[3] *Principles of Appraisal Practice and Code of Ethics,* American Society of Appraisers, Washington, DC. Jan. 1994, p. 3. Language not applicable to the subject at hand has been edited out.

[4] The appraisal of "certain other business assets" is referred to in this book as "technical specialties." See chapter 11 for more information on technical specialties.

[5] USPAP, p. 11.

[6] Fair market value can sometimes include removal costs, such as in situations in which it is typical for buyers to adjust amounts offered to offset any removal costs.

[7] The appraiser has two options for responding to the assumption that business earnings support the value reported for the underlying asset: one is to assume, without verification, that this is the case; the other is to actually determine whether there are sufficient earnings by using an income approach. If the first option is selected, the appraiser must ensure that the appraisal report clearly states that the value reported for the underlying asset assumes, without any verification, that business earnings are sufficient to support the value conclusion.

[8] The term *auction* usually refers to forced liquidation value, but there are exceptions to this general rule; for example, in certain industries, an auction is the standard industry method for disposing of assets, in which case it may be equal to orderly liquidation value, assuming a normal exposure time (and may be equal to fair market value under certain conditions). The essential difference between orderly liquidation value and forced liquidation value is one of exposure time.

[9] *DREA,* p. 234.

[10] *The Appraisal of Real Estate (ARE),* p. 19. Appraisal Institute, Chicago, 11th ed., 1996.

[11] *ARE,* p. 19.

[12] *DREA,* p. 72.

[13] For an alternative viewpoint, see Henry A. Babcock, *Appraisal Principles and Procedures* (Washington, DC.: American Society of Appraisers).

[14] *ARE,* p. 336; James C. Bonbright, *The Valuation of Property* (New York: McGraw-Hill, 1965), p. 157.

[15] Eugene L. Grant and Paul T. Norton, Jr. *Depreciation* (New York: The Ronald Press Company, 1955), p. 269.

[16] Throughout this book, the property to be appraised is often referred to as the "subject property."

[17] American Institute of Certified Public Accounts (AICPA), Accounting Principles Board. *Opinions* (New York: AICPA).

2

Classification and Description of Machinery and Equipment

Objectives:

1. Define appraisal *accounts* and *classes*.
2. List typical machinery and equipment classes.
3. Explain the process of listing and describing machinery and equipment.
4. Discuss sampling and desktop techniques.

Experienced machinery and equipment appraisers are professional fact and data collectors. They are keen observers and resourceful investigators who are not likely to be intimidated by an occasional lack of precise knowledge of the process or machines they are called upon to appraise. They know that if their *classifications* and *descriptions* of the subject property are good, they will have the information they need to perform the required valuation research and reach supportable conclusions of value.[1]

This chapter discusses the classification and description (often called "listing") of machinery and equipment. With a few exceptions, the basic techniques for classifying and describing assets are similar whether the appraiser is valuing an entire plant or a single machine and regardless of the industry or process being appraised.[2] This chapter discusses the listing of entire factories as well as of single machines. Large steel mills, whose production is often influenced by global developments, are usually self-contained units. A complete steel mill may have a railroad, a power generation plant, power distribution lines, water treatment facilities, quality control laboratories, production facilities, research facilities, material handling equipment, and administrative offices. Despite this wide array of assets, the appraiser uses the same basic classification and description methods to list the assets of the steel mill that he or she would use to list the assets of a single-chair dentist's office.

Individual appraisers have various styles of assembling information. Listing a massive process plant can be relatively simple if approached in an organized manner. Simplification is a major concern to the appraiser, plant owner, and—if one is involved—lender. Of course, simplification is no excuse for insufficient detail. The other extreme in description is verbosity. Seasoned appraisers learn from training and actual experience what information should be recorded to accomplish their objectives.[3]

Before listing the assets, it is worth reviewing the extent of the engagement. An important consideration is the scope of the appraisal work: are you appraising the entire plant, only a specific portion of the plant (e.g., a process line), or individual machines?[4]

Looking at the flowline in figure 2.1, note how the illustration has simplified a complex process into small "visuals." Equipment or process descriptions, like graphic visuals, are important to the appraiser and client. For example, captions at the left-hand corner read "Iron Ore" with the flow toward "Pellets." Intuitively one recognizes that the iron ore and pellets receive more processing than the drawing illustrates. The appraiser also understands, without a long explanation, the intrinsic value of the machinery and knows that a business enterprise developed the engineering, assembled the labor, found the iron ore deposit, purchased the machines to mine the deposit, transported the ore, and processed it in other machines to get to the next step of pelletizing. Whenever a manufacturer produces material, moves the product, processes the product, handles the product, or in any way changes the product, it is adding value to the product. If the product has value through each stage of processing, then the machinery that processes and adds value must also have value. Each stage in the handling and changing of the product is a small plant or system. Each component requires an initial purchase and turnkey costs, such as freight and installation, to make the machine a producer.

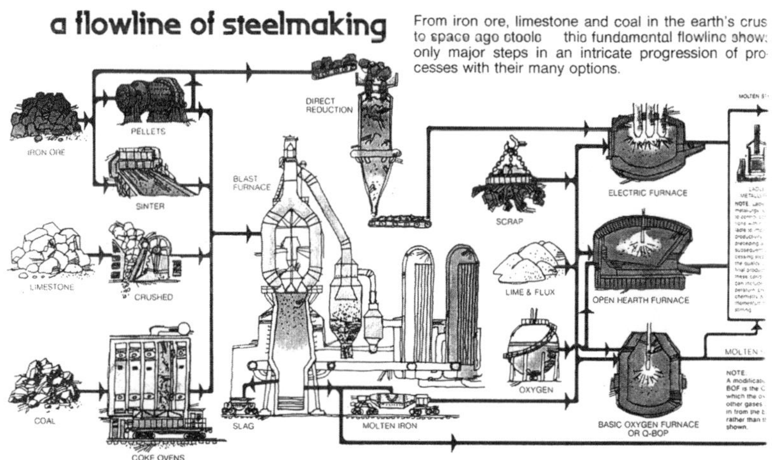

Figure 2.1. Flowline of steelmaking. *Courtesy of the American Iron and Steel Institute.*

Referring again to the visual aspect of description, the experienced appraiser will think of a manufacturing plant as a group of individual items brought together and assembled to generate a financial return. As the appraiser inspects and lists each machine, it is considered a "visual"; that is, the very fact that the machine can be seen says something about its use, capacity, and relationship to other machinery in the area. After developing an overview of the facility, the appraiser's next task is to find, identify, classify, describe, and ultimately value each component relative to its physical and economic condition and its market supply and demand.

Accounts and Classes

A standardized group of appraisal accounts and classes promotes uniformity and consistency. For appraisal purposes, "accounts" are major groupings of assets that are similar in character. The most basic separation of tangible assets into accounts would be land, buildings (or structures), and equipment (including machinery and furniture, although some would classify furniture as a separate account).[5]

The equipment account can be broken down further into various "classes," such as production machinery, general plant equipment, office furniture and fixtures, and other classes (described in the next section). Using the client's accounts and classes is often desirable, especially when the appraiser's work must be reconciled to the client's accounting records. In such cases, the client's equipment classes may differ from those described in the following text, although they will probably be similar.

Typical Machinery and Equipment Classes

Production Machinery

Production machinery is usually the appraiser's primary concern because it usually represents a relatively large percentage of the total value of a plant (see figure 2.2). Anything other than the main production machinery or process is peripheral to, and in support of, the operation. For example, the main part of a cement manufacturing plant is the lime kiln. The electrical motor control center (MCC) is a support item (of course, without the MCC the kiln and the entire plant are inoperable). It is essential for the appraiser to understand the manufacturing process. Although industry research is not discussed in this chapter, we will assume that the appraiser, starting a new and unfamiliar plant evaluation, will have researched the process through library or industry sources. At the very least, the appraiser should review the process with the appropriate plant personnel before beginning the listing process.

Figure 2.2. Example of production machinery. In large operations, some of these production machines may actually act as support equipment to a complete process.

Support Equipment

Machinery that maintains or increases the ability and capacity of the production equipment is support equipment (see figure 2.3). For example, most plastic injection molding machines have a scrap grinder associated with them. The injection machine can process plastic without the grinder, but the injection process is less efficient. The scrap grinder grinds or cuts the excess plastic for reuse.

CUMBERLAND LEESONA
SIZE 484-GRAN-3KN 14" X 16"
PLASTICGRANULATOR

Figure 2.3. Example of a support machine.

Motor Control Centers and Switchgear

Motor control centers and switchgear are often referred to by the acronym MCC. Unless the plant has main electrical substations, the MCC is usually the largest electrical distribution center in the plant. The MCC is electrically fed by transformers, usually outside the building, which in turn are fed by high-power lines or substations. There may be MCCs for each area, or one MCC may control the electrical power for the entire plant. The MCC can vary from a few wall-mounted main switches to a major installation including a separate air-conditioned enclosure (see figure 2.4). Primary electrical voltage is measured in volts, amperes, and watts.

Electrical current is either alternating current (AC) or direct current (DC). The most common industrial electricity is AC, which is cycled as either 50 or 60 hertz. Power required for machines is measured in horsepower (HP) or kilowatt (kW). Electrical current is measured in amperes (amps). Industrial power is usually referred to as three-phase. Most equipment under 1 HP is run on single-phase electricity. Transformed voltages to the industrial motors are often 208-230/460 volts AC (in Canada, 550 volts AC). Exceptions occur, depending on the motor manufacturer specifications and the available utility service. In some countries, including much of Europe, transformed electricity may be 380 volts. Despite the differences in voltage and hertz, it is not unusual to find mixed voltages in the same plant. The uncommon voltage will have been transformed, rectified to the plant incoming power, or run in an alternate hertz.

MCC areas can be deceptive in that the single-phase power may be installed in the same area as the plant's three-phase power. Although single-phase power is used to run small equipment, usually less than 1 horsepower (HP), it is often considered part of the wiring of the building. Therefore, much of the single-phase power may not be considered machinery unless the processing or manufacturing equipment is single-phase power. For example, laboratories, medical offices, electronics, and small manufacturing lines use extensive single-phase power of less than 1 HP.

It is important to consider the MCC when appraising the plant as a whole or appraising under some form of *fair market value in continued use* or *fair market value installed.* The MCC may have less relevance in estimating a distressed sale value when the equipment is to be removed.

Figure 2.4. Different types of motor control centers.

Power Wiring

The plant power wiring connects the plant equipment to the MCC. The equipment is strung together with wiring, controls, and switches. If the appraisal requires a listing of the power wiring, the appraiser will estimate the size, number of lineal feet of conduit and wire, number of motors, supply boxes, and switching between the equipment and the MCC.

Process Piping

Many plants require compressed air, water, steam, gases, or fluids for heat processing or cooling. For example, papermaking, textile dying, and ore processing require large amounts of water. Process piping (see figure 2.5) can include piping, valves, fittings, pumps, and process controls, among others, and is important in appraisals for *fair market value in continued use* but less important in liquidation appraisals.

LANCY INTERNATIONAL WASTE TREATMENT SYSTEM

PARTIAL VIEW OF LANCY INTERNATIONAL WASTE TREATMENT SYSTEM AND CONTINENTAL WESTERN RODI WATER SYSTEM

Figure 2.5. Examples of process piping.

Foundations and Structural Supports

A valuation under a *fair market value in continued use* premise generally requires that the appraiser include associated costs for equipment installation. Some massive machinery requires special pits or heavy foundations (see figure 2.6). Installed machinery may require structural steel, catwalks, ladders, or platforms. The appraiser may or may not include these related structures, again depending on the scope of the assignment. For example, in North America, most insurance-related appraisals would not include anything below the lowest concrete floor of the plant. On the other hand, an appraisal for allocation of purchase price may require the inclusion of all the costs required to bring a machine into production. Regularly published guides and books will aid in the description and pricing of these items. Some of these guides will be discussed later in the text.

Figure 2.6. A machine installation with foundations and structural supports.

Material Handling and Storage Equipment

Plant production depends on moving materials through the plant, from raw material receiving to material processing, product finishing, and inventory. This is accomplished with material handling equipment such as forklifts, loaders, conveyors, cranes, hoists, pallet movers, programmable inventory systems, and the like.

General Plant Equipment

General plant equipment includes mainly lower unit cost items necessary for the operation of the plant, such as plant furniture and fixtures (located in manufacturing instead of office areas), benches, racks, lockers, scales, hand trucks, time recorders, ladders, fire extinguishers, and similar assets.

Rolling Stock

Some appraisers break rolling stock into its two subclasses: plant vehicles and licensed vehicles. "Plant vehicles" include trucks, cranes, tractors, and other mobile equipment not licensed for road use. "Licensed vehicles" include automobiles, trucks, tractors, trailers, and other vehicles licensed for road use. Depending on the type of operation, mobile equipment can be of great or little consequence. For example, in mining operations, the mobile equipment is generally a significant portion of the total value. Most manufacturing plants have delivery trucks and some automobiles. Construction companies use material-handling trailers, loaders, tractors, and other mobile equipment. Some appraisals are limited to the rolling stock itself.

Laboratory and Test Equipment

Laboratory and test equipment include items necessary for the operation of a laboratory or test facility. Typical equipment would include microscopes, clean tables, fume hoods, ventilating systems, spectrographs, ovens, stills, glassware, and similar assets.

Office Furniture, Fixtures, and Equipment

The appraisal may include office furniture, fixtures, and equipment. This class includes desks, tables, chairs, credenzas, filing cabinets, portable partitions, calculators, copy machines, fax machines, check writers, typewriters; and similar assets. Computers (personal, network, and mainframe), monitors, printers, plotters, modems, scanners, and similar assets may be included in this class, or they may be included in a separate class (see the following section). Some clients only want items listed that have value over a certain amount, with assets under a specified value being aggregated. The appraiser must use caution in these situations so that the client's requests do not make the appraisal misleading.

Computer Equipment

This class includes various computers (personal, network, and mainframe), monitors, printers, plotters, modems, scanners, and similar assets, if they are not included in "office furniture, fixtures, and equipment." It may also include computer-aided design (CAD) and computer-aided manufacturing assets that support manufacturing operations. Many manufacturing systems are "hardwired" between the engineering, accounting, and the manufacturing areas. Office systems are networked between personal computers and fileservers. Manufacturing facilities and research departments are often connected to the Internet or other communications systems. An appraiser may be asked to value not only the hardware but sometimes the software associated with these systems.

Tools

This class is often divided into two subclasses: permanent and perishable. "Permanent tools" includes portable electric and air tools, anvils, vises, gauges, chucks, and similar items. "Perishable tools" includes items that are consumed within a relatively short period, such as drills, chisels, reamers, taps, and other assets.

Special Tooling

This class is sometimes called "tools, dies, jigs, and fixtures." It includes apparatuses used in conjunction with production equipment and built for specific operations or applications. Items in this category include dies, jigs, fixtures, molds, patterns, templates, and similar assets.

Construction in Progress

Construction in progress includes projects under construction that have not been completed and capitalized.

Special Classes

Appraisals for certain industries may require additional classes, such as aircraft, bottles and cases (beverage industry), screens (printing), cores and molds (foundry), and trays and pans (bakery). At other times, it may be necessary to augment the aforementioned classes to account for the following situations:

Leased: It is common to find leased equipment used in combination with the company's owned assets. Leases may include office equipment or even an entire manufacturing line. It is important to identify leased equipment. The appraisal report should contain a limiting condition stating the appraiser's reliance on the client's representations regarding leased equipment.

Not inspected or away from premises: This situation refers to assets that were not inspected for a valid reason, for example, assets located in a remote or hazardous area, in which case it might be reasonable to rely on the client's information regarding description, condition, and so forth (the appraisal report should be qualified accordingly).

Nonoperating assets: This situation refers to assets that are extraneous to business operations in the sense that they do not contribute to the earning capacity of the business entity. These assets are often held for purposes of investment, subsequent disposal as surplus property, or other reasons.

Inventories

Often, the appraiser will not be required to appraise inventories. However, there are appraisers who specialize in the valuation of inventory. Inventories are typically appraised in connection with factoring or some type of loan (and they are sometimes appraised when a purchase price needs to be allocated). Factoring is a method of financing seasonal inventory such as a summer line of clothing or seasonable ornaments. Factoring is also used to complete current contracts or borrow money based on accounts receivable. In manufacturing, inventories can generally include raw materials, work-in-process, finished goods, and dead stock.

Identification and Listing of Machinery and Equipment

There are two major procedures in the identification and listing of machinery: macroidentification and microidentification.[6]

Macroidentification

Macroidentification studies the entire manufacturing process by identifying major components contributing to the design capacity of the plant.[7] Macroidentification is the method the appraiser uses to answer these questions: (1) What does the plant produce? (2) How is the product produced? (3) What is the capacity of the plant?[8]

Information to be considered when gathering the data for macroidentification of machinery and equipment is given in the following list:

- Date
- Company name and address
- Source of information
- Products produced (with each process name and description)
- Engineering design firm and contractor
- Original date of construction and expansions
- Plant or process byproducts, amounts, and uses
- Plant or unit capacity per day or per year (tons per day, gallons per day, barrels per day, annual production, etc.)
- Plant capacities: design capacity; rated and actual consistent capacity
- Plant efficiency or obsolescence
- Yield or losses and reason for losses
- Feedstocks and sources
- Operating mode (days, month) if not identified in capacity (e.g., sugar beet plants that run seasonally)

- Outlets for finished or intermediate products
- Plant sales outside the parent company for use in other company plants, or other product sales possibilities
- Available historical operational data over three to five years
- Fuel and power consumption by unit
- Operating staff per unit
- Type of control systems and if the control is centralized
- Estimated maintenance budgets over the past three to five years and projected budget if plant is operational
- Identification of equipment requiring extraordinary maintenance, with reasons given
- Maintenance program implementation (i.e., regular, preventive, or demand)
- Plant layout: Is the flow considered adequate, manageable, etc.?
- General condition of plant and components
- Age: chronological, effective, and estimated equipment, and product remaining useful life
- Status of safety and environmental standards: Are they good, and if not, can they be upgraded, and at what cost?
- Pollution control equipment in place
- Support facilities
- Obvious detrimental factors[9]

Obviously, not all of these items apply to all appraisals, and the appraiser should exercise judgment regarding the information that needs to be obtained.

Figure 2.7 illustrates an abbreviated example of the type of information that may be pertinent in the macroidentification process. The depth of investigation is contingent on the scope of the appraisal, that is, whether the appraisal is for a group of machines, a technical process, or a system.

One of the most important items the appraiser identifies is the capacity—that is, what was the line designed to process, what is its designed or rated output, and what is the current or actual output? For example, an appraiser might include the following information in his or her description of the coil line shown in figure 2.7:

General Description

A steel coil to sheet line for the production of 126" wide × 1" thick mild carbon steel and hot-strip low-alloy steel at 100,000 psi tensile strength and 75,000-lb. yield.

Coil Size

Maximum weight:	100,000 lb.
Inside diameter:	30"
Width:	43" to 126"
Thickness:	3/16" to 1"
Pounds per inch of width:	250 lb. to 1500 lb.
Coil temperature:	–10°F to 110°F

Speeds: 0 to 250 feet per minute (FPM); 250 FPM maximum for mild steel to 0.500" and up to 150 FPM for heavier gauges up to 1.000".

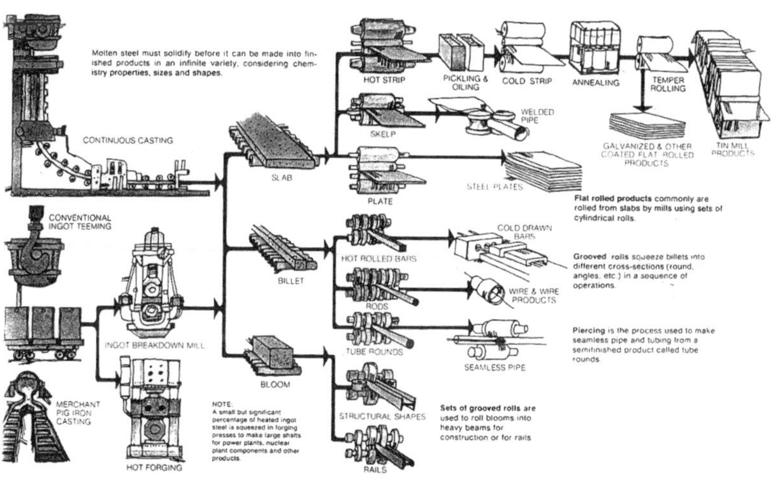

Figure 2.7. Continuous caster to finishing lines.
Courtesy of the American Iron and Steel Institute.

This kind of information may be obtained from the plant engineer or from process drawings. On the other hand, if the appraisal is of a small shop or other process where there are little data, it may be necessary to interview management to obtain the information. The primary focus is to find out what the plant manufactures, how much, and over what period. Once these questions have been answered, the machinery and peripheral equipment can be listed accordingly.

Microidentification

Microidentification is the process of finding the individual characteristics of the equipment; it focuses on the listing of a single machine and identifies the specifics of the equipment. Of prime importance in microidentification is the brand name, model number, serial number, type of power, and dimensions (if practical):[10]

Brand name. The brand name or manufacturer may be the only identification that can be found. Occasionally, there will be no identification because the nameplate is missing or cannot be read.

Model number and size. The model number and size may be the same, or the model number may have no relationship to size but merely represents the manufacturer's cataloging number.

Serial number. The serial number can be an extremely important piece of information. Given a serial number, a call to the manufacturer might result in the determination of the year of manufacture, capacity, size, and sometimes even the original owner. For some machinery, such as machine tools and construction equipment, the data are so useful that serial number guides are regularly published.

Capacity. Capacity is often identified in the size of a machine. For example, a backhoe tractor-loader is often measured by how many yards or cubic meters it can dig; a crane is measured by the number of tons it can effectively lift; and a milling machine is measured by the horsepower or kilowatts and the worktable size.

Physical size. Sometimes it is useful to measure the physical size or footprint of the equipment—for example, the length of a conveyor, the diameter and height of a tank, or the span and travel of a crane.

Age and condition. Chronological age can often be obtained from the manufacturer's plate or serial number. Condition is a subjective observation about the physical appearance, and sometimes the operating performance, of the machine.

The following is an example of how an appraiser might record microidentification information for the shear shown in figure 2.8:

Item:	Steel plate shear
Manufacturer:	Cincinnati
Model:	1410
Serial number:	32017
General description:	Power squaring shear, 10' × 3/16"; 3/16" maximum shearing capacity, 8' squaring arm; front-operated power back gauge
Motor:	7.5-HP main motor 220/440/3/60
Condition:	Good

Figure 2.8. Steel plate shear.

For a *fair market value in continued use* or *fair market value installed* appraisal, the appraiser normally includes several other items in the description, such as foundations, structural supports, wiring, and switches. These items are usually excluded if the appraisal is for liquidation.

During the initial inspection of the assets, it is important to record as much detail as reasonably possible. As an appraiser's experience grows, so does his or her ease in gathering data. Since information can always be refined after the appraiser has all the facts, it is usually better to err on the side of collecting too much detail. However, there is a point at which additional detail may be superfluous or even counterproductive. For example, if the engagement is a large one, if there are staffing restraints, or if the client needs a very quick turnaround, the collection of unnecessary details should be avoided. Listing essential information such as manufacturer, model number, serial number, capacity, and condition usually provides sufficient information to complete the necessary valuation research.

Given that individual machines are usually components of larger processes, another method of quick microidentification is to list the entire line as a plant component. Take, for example, a complete fluid milk pasteurizing line. Typically, these processes include a feed storage tank, a homogenizer, a high-temperature-short-time pasteurizer, holding tubes, balance tank, pumps, controls, valves, and sanitary product transfer tubing. One way to appraise this line would be component by component. However, the appraiser could also identify and appraise the same process by capacity. For example, the milk plant can be measured in pounds per day or in gallons per hour. On the basis of this information, an appraiser could use engineering data to identify the equipment and price the components. Similarly, other types of plants, such as cement plants, certain mining operations, and some steel production facilities, can be valued by units of production. Sometimes, the unit cost method is the most practical method of identification and valuation, especially when there are ample industry data to support its use. With proper statistical support, this method can be used for technical valuation in property tax matters, business analysis, public utility studies, right-of-way, and public domain cases.

Identification through Sampling and Desktop Analysis

Sampling

Sampling is a valuation technique that may be used to appraise large amounts of tangible assets without actually inspecting every item. The sampling process might check one of a dozen items, 10 percent of the total equipment cost, or several assets in each equipment class. Sampling is generally appropriate only when an accurate reflection of the whole can be inferred from the sample. The problem with sampling is that the appraiser does not see everything and must make certain assumptions regarding the assets he or she has not seen. Before using the sampling technique, the appraiser should explain to the client the limitations of this form of appraisal, obtain the client's agreement to use the method, properly disclose the method in the appraisal report, and satisfy any other requirements of USPAP.

Desktop Analysis

Another identification method is the desktop analysis, sometimes called the "examination" or "audit" method. The desktop analysis identifies and values assets without inspecting and listing them, by relying on documents and other information provided by the client or others. This method can be useful, especially when there are large amounts of equipment in multiple locations and little time to complete the appraisal. However, it should be remembered that both the desktop and sampling techniques are a form of "limited appraisal," which is defined by USPAP as "an estimate of value performed under and resulting from invoking the Departure Provision." Both methods result in something less than that normally required by USPAP; thus, the appraiser should be careful to comply with applicable ethical requirements and should satisfy himself that the use of these abbreviated methods will not result in a false or misleading appraisal.

Key Points

- Experienced machinery and equipment appraisers are professional fact and data collectors. They are keen observers and resourceful investigators who are not likely to be intimidated by an occasional lack of precise knowledge of the process or machines they are called upon to appraise. They know that if their *classifications* and *descriptions* of the subject property are good, they will have the information they need to perform the required valuation research and reach supportable conclusions of value. For appraisal purposes, *accounts* are major groupings of assets that are similar in character. The most basic separation of tangible assets into accounts would be land, buildings (or structures), and equipment.

- The appraiser breaks the equipment *account* down into various *classes*. Typical equipment classes are production machinery;

support equipment; motor control centers and switchgear; power wiring; process piping; foundations and structural supports; material handling and storage equipment; general plant equipment; plant and motor vehicles; laboratory and test equipment; office furniture, fixtures, and equipment; computer equipment; tools; special tooling; patterns and templates; construction in progress; special classes; and inventory.

- Two major procedures in the identification and listing of machinery are *macroidentification* and *microidentification.* Macroidentification is the study of the entire manufacturing process by identifying major components contributing to the design capacity of the plant. Microidentification is the process of finding the individual characteristics of the equipment; it focuses on the listing of individual machines and identifies the specifics of the equipment. Of prime importance in microidentification are the brand name, model number, serial number, type of power, and dimensions (if practical).

- *Sampling* and *desktop analysis* are *limited appraisal* methods that may be used to value large amounts of inventories or other fungible assets without actually inspecting each item. The appraiser should remember that these methods result in something less than that normally required by USPAP. Thus, the appraiser must comply with all applicable ethical requirements and should satisfy himself that use of these abbreviated methods will not result in a false or misleading appraisal.

Additional Reading

American Society of Appraisers. *Appraising Machinery and Equipment.* Washington, DC: McGraw-Hill, 1989.

ASA Boston Chapter, Machinery and Equipment Section, *Industrial Properties/Machinery and Equipment Study Guide,* from 1976 appraisal conference, Boston, 1976.

Byrns, Ralph T., and Gerald W. Stone. *Economics.* 3rd ed. Glenview, IL: Scott, Foresman, 1987.

Denver Equipment Company, *Modern Mineral Processing Flow Sheets*, 2d ed., Colorado Springs, CO, 1979.

Heldman, Dennis R., and R. Paul Singh, *Food Process Engineering*, 2d ed., AVI Publishing, Westport, CT, 1981.

Jackson, J. M., and Shinn, B. M., *Fundamentals of Food Canning Technology*, AVI Publishing, Westport, CT, 1979.

Kharbanda, O. P., *Process Plant and Equipment Cost Information*, Craftsman, Solana Beach, CA, 1979.

Leffler, William L., *Petroleum Refining for the Non-Technical Person*, Pennwell, Tulsa, OK, 1979.

McGinnis, R. A., *Beet Sugar Technology*, 3rd ed., Beet Sugar Development Foundation, Ft. Collins, CO, 1982.

Meade, William, P. E. (Ed.), *Encyclopedia of Chemical Process Equipment*, Reinhold, NY, 1964.

Merkel, James A., Ph.D., *Basic Engineering Principles*, 2nd ed., AVI Publishing, Westport, CT, 1983.

Modern Plastics Encyclopedia, McGraw-Hill, NY, yearly update, 1985–1986.

Perry, J. H., C. H. Chilton, and S. D. Kirkpatrick, *Perry's Chemical Engineer's Handbook*, 4th ed., McGraw-Hill, NY, 1963.

Potter, Norman N., Ph.D., *Food Science*, AVI Publishing, Westport, CT, 1968.

Pride, William M., and O. C. Ferrell. *Marketing: Basic Concepts and Decisions.* 5th ed. Boston: Houghton Mifflin, 1987.

Short, J. A., *Drilling, A Source Book on Oil and Gas Well Drilling from Exploration to Completion*, Pennwell, Tulsa, OK, 1983.

Slate, L.D., *How to Buy Metalworking Machinery*, Hearst Business Media Corp., IMN Division, Southfield, MI, 1977.

Thomas, R. (ed.), *Engineering and Mining Journal Library of Operating Handbooks*, vol. I, *Handbook of Mineral Processing*, McGraw-Hill, NY, 1977.

Notes

[1] Alan C. Iannacito, "Identification of Machinery and Equipment," in *Appraising Machinery and Equipment* (Washington, DC: American Society of Appraisers, 1989). p. 27.

[2] Ibid., p. 17.

[3] Ibid., p. 18.

[4] Ibid., p. 17.

[5] Skousen, K. Fred; Harold Q. Langenderfer; and W. Steve Albrecht, *Financial Accounting,* 4th ed. (Cincinnati, OH: South-Western Publishing Co., 1991), p. 314.

[6] Iannacito, "Identification of Machinery and Equipment," pp. 18–26.

[7] Ibid., p. 18.

[8] Ibid., pp. 21–22.

[9] Ibid., pp. 22–23.

[10] Ibid., p. 23.

3

Cost Approach

Objectives:

1. Introduce the cost approach and principle of substitution.
2. Distinguish between replacement and reproduction cost.
3. Describe methods of determining cost new.
4. Discuss concepts of depreciation, including obsolescence.
5. Illustrate the cost approach methodology with examples.

Using the cost approach, the appraiser starts with the current replacement cost of the property being appraised and then deducts for the loss in value caused by physical deterioration, functional obsolescence, and economic obsolescence. The logic behind the cost approach is the principle of substitution: a prudent buyer will not pay more for a property than the cost of acquiring a substitute property of equivalent utility.[1] The principle can be applied either to an individual asset or to an entire facility.

In its simplest form, the cost approach is the current cost (as if new) less all depreciation. The appraiser identifies the property being appraised ("subject"), develops its current replacement cost new, and subtracts all depreciation that makes it less desirable to own than if it were new.

Determination of Current Cost New

Replacement and Reproduction Cost

The replacement cost new is generally the proper starting point for developing an opinion of value using the cost approach.[2] It is essential that the appraiser understand the difference between *replacement cost new* and *reproduction cost new*. Replacement cost is the current cost of a similar new property having the nearest equivalent utility as the property being appraised, whereas reproduction cost is the current cost of reproducing a new replica of the property being appraised using the same, or closely similar, materials. In using the cost approach, the appraiser is comparing to the subject property the property that could actually replace it. The replacement property would be the most economical new property that could replace the service provided by the subject.[3] As Professor Bonbright states:

> Most physical properties are not replaced by properties of the same size, design, and materials. They are replaced by materially different properties of a more modern type, better designed to meet the owner's present needs.... [T]he replacement would be one of substitution [i.e., replacement cost] rather than identical reproduction.... In such cases, the hypothesis that the value of the existing property is derivable from the current cost of constructing or buying a substantially identical property [reproduction cost] is always invalid. The appraiser may still adhere to it [reproduction cost] if he believes that there is no material difference between the cost and efficiency of the different substitute [replacement cost] and the cost and efficiency of the replica [reproduction cost]. But he cannot ignore the discrepancy if it is serious—otherwise he will be guilty of gross overvaluation.[4]

Professor Grant makes the same point: "[I]f replacement with a different substitute asset would be more economical, the cost of reproducing identically an existing old asset has no bearing on value."[5]

Whether the subject asset is an individual item of equipment or an entire plant, improvements in design, product flow, processing methods, layout, and equipment size and mix make the modern substitute more desirable to own. It will be more desirable to own because it is superior to the subject in one or more of the following respects: it costs less to acquire (capital cost advantage), costs less to operate (operating cost advantage), or produces more revenue (profit advantage).

Although replacement cost is the proper starting point in the cost approach, this does not preclude development of reproduction cost for some purposes. Reproduction cost can be developed to quantify one form of functional obsolescence, that due to *excess capital costs*. Assuming that a property's replacement cost is less than its reproduction cost,[6] *excess capital cost* is measured by the difference between reproduction and replacement cost. This excess cost represents the lower capital investment required to obtain the most economical new asset to perform the same service as the subject.

Reproduction cost can also be estimated when an appraisal is to be compared with other value indicators developed by trending, so that there will be an equal basis of comparison (i.e., an "apples to apples" comparison). In addition, if the subject asset would actually be replaced with a replica, there is no difference between replacement and reproduction cost and, hence, there is no harm in referring to the starting point as reproduction cost new (although the term *replacement cost new* would also be correct).

The difference between replacement and reproduction cost has been discussed above in theoretical or definitional terms. The following example applies the theory to a hypothetical set of facts.

Example 1: Replacement or reproduction cost estimated by using the proper level of current cost as the starting point.

The task is to appraise a 50,000-barrel-per-day petrochemical refinery and your role is to appraise only the steam generation facilities. You take a plantwide tour and ascertain that the steam generated is for process use only. You meet with the boiler superintendent to discuss the facilities. In your meeting, you establish there are four boilers and you receive the information presented in table 3.1.

Boiler Number	Rated Capacity (Pound Per Hour)	Chronological Age in Years	Furnace Type
1	15,000	46	Brick-set oil fired
2	30,000	36	Field erected oil-fired
3	30,000	10	Package oil-fired
4	60,000	6	Package oil-fired
Total	135,000		

Table 3.1. Example 1

Other Information:

Boiler 1 is still capable of operating but has been considered excess since boiler 4 began operation six years ago. Boiler 1 has not operated since boiler 4 was installed.

Boiler 2 is a "swing" boiler and runs only occasionally during peak periods.

Boiler 3 operates in conjunction with boiler 4 and usually operates at or near capacity.

Boiler 4 runs continuously, which maximizes its use.

Peak periods are midsummer and midwinter.

a. Peak demand in summer is 100,000 pounds per hour.

b. Peak demand in winter is 90,000 pounds per hour.

The client has recently built a modern refinery of equal capacity to the subject in a neighboring state. You contact the manager of that refinery to establish the characteristics (cost, design, engineering, technology, construction materials, etc.) of the modern facility that would, hypothetically, replace the subject facility. You are provided with the following information concerning the company's new refinery:

The new refinery is of equal capacity, i.e., 50,000 barrels per day.

Construction of the new plant was completed three years ago.

The new plant uses heat recovery on the process units to reduce plant-generated steam; consequently, peak demands in summer and winter have virtually been eliminated in the new plant (the subject refinery does not have a heat recovery system).

The new plant has two 40,000-pounds-per-hour package-type boilers, oil fired, for a total rated capacity of 80,000 pounds per hour.

The historical cost of the new refinery's boilers, including all peripherals, was $1,500,000 (three years ago).

The question in this example is "What is the appropriate current cost new to be used in the cost approach?" Assume the following: The purpose of the appraisal is for property tax purposes (to determine whether the assessed value placed on the property by the local assessor is fair and reasonable); the premise of value is fair market value in continued use; all costs are midyear; and the trend factor is 10 percent[7] per year.

Begin by analyzing the available facts. Given that the company just built a new refinery, you have the basis for establishing the replacement cost new for the entire refinery, and specifically, the steam generation facilities. (Assume that the feedstock and product mix of the new and subject facilities are the same.) When you look at the subject boilers, you see a number of areas of obsolescence. Boiler 1 is brick-set/oil-fired and represents excess construction. Although it is not mentioned specifically, it would be logical to assume that the use of heat recovery on the refinery process units represents an advance in technology resulting in reduced steam requirements. The subject plant has a total steam capacity of 135,000 pounds per hour, while the new plant has a total of 80,000 pounds per hour, reflecting excess capacity in the subject plant. Since a prudent purchaser would pay no more than the cost of acquiring the most economical new asset that could replace the service provided by the subject, the appropriate current cost would be the cost of two 40,000-pounds-per-hour package-type, oil-fired boilers. That cost would be calculated as follows: the historical cost (three years ago) of \$1,500,000 is multiplied by a trend factor of 1.331 (1.10 × 1.10 × 1.10) to bring the historical cost of the modern replacement asset to the current date; thus, \$1,500,000 × 1.331 = \$1,966,500. In this example, the proper level of current cost would be a *replacement cost new* of \$1,966,500.

Although the *reproduction cost new* of the four subject boilers will not be calculated here, it should be noted that the replacement cost of \$1,966,500 for the modern replacement asset is significantly less than the reproduction cost of the four existing boilers of varying sizes and types.

In this example, a minor problem needs to be discussed. The issue of plant-generated steam, which has been reduced in the hypothetical replacement plant by using heat recovery on the process units, has not been fully addressed. Although there is a corresponding cost savings for the steam generation facilities, there is probably an increased capital cost associated with the additional equipment needed on the process units to recover waste heat. The point is that the replacement plant needs to be balanced from both a capital and operating cost standpoint. This is especially true when appraising integrated facilities such as oil refineries or other process plants.

Replacement cost is often, but not always, less than reproduction cost. The following example presents a situation in which replacement cost is greater than reproduction cost. The question arises: Why would an owner replace an existing property (i.e., the subject) with a property the replacement cost of which is greater than the reproduction cost of the subject? This example begins to explain the answer, although the full explanation may not be apparent until functional obsolescence is discussed.

Example 2: Replacement cost greater than reproduction cost.

The task is to appraise a machine that we will call Model A, using the following information:

Chronological age of Model A is six years.

Model A is no longer manufactured and has been replaced by Model B.

The last selling price of a new Model A was $100,000.

The reproduction cost new of Model A (determined by trending) is $130,000.

The capacity of both models is 100 units per day.

At first glance, it might be tempting to use the reproduction cost new as the starting point of the cost approach. However, note that Model A is no longer manufactured and has been replaced by Model B. After further investigation, the appraiser determines that Model B also has a capacity of 100 units per day, but its current cost is $160,000, which is higher than the reproduction cost new of Model A. The reason for the capital cost difference is an advance in technology that significantly reduces the energy required to produce each unit. On an annual basis, the subject (Model A) requires 100,000 therms (one therm = 100,000 BTU) of natural gas, whereas Model B requires only 75,000 therms, a difference of 25,000 therms. Assuming 50 cents per therm, Model B will save $12,500 per year in operating cost, which will justify purchasing Model B at the higher price.

In this example, the new model is more desirable than the older one because it can replace the service provided by the subject (i.e., it can produce an equivalent number of units per day) with reduced energy consumption, which means lower operating costs. The initial capital cost of the new model is greater than the reproduction cost of the older model, but the new model more than makes up the difference in reduced operating costs. Therefore, even though the replacement cost of Model B is greater than the reproduction cost of Model A, the proper starting point for the cost approach would be the replacement cost new of Model B, or $160,000. Later, when we discuss functional or operating obsolescence, we will measure the additional depreciation attributable to the subject property (Model A) resulting from the appraiser's comparison of the subject to the new model with its reduced operating cost (see example 11).

Methods of Determining Cost New

There are several methods of determining the current new cost of a property. The major ones are the *detail method*, *trending*, *cost-to-capacity*, and *other engineering methods*.

Detail Method

The *detail method,* also known as the summation method, requires that a current new cost be assigned to each individual component of an asset or property. The property is itemized or "detailed" so that the sum of the components reflects the cost new of the whole.

All normal or typical direct and indirect costs should be included. *Direct costs* are those material, labor, and related expenditures normally and directly incurred in the purchase and installation of an asset, or group of assets, into functional use. Examples of direct costs are

Direct material costs, including the item of equipment itself

Direct labor costs for installation or erection of the item

Freight and handling

Rigging and moving

Electrical

Piping

Foundations

Millwrighting

Sales tax

Indirect costs are those expenditures that are normally required to purchase and install a property but which are not usually included in the vendor invoice.[8] Examples of indirect costs include the following:

Engineering, architect, and other professional fees

Administrative, accounting, consulting, and legal fees

Temporary insurance during installation or construction

Licenses, permits, and fees for installation or construction

Security costs during installation or construction

Typical finance charges during installation or construction

Equipment rental

Temporary enclosures

Debugging and run-in costs

When developing cost new, only those direct and indirect costs that are typical or normal are included; unusual, atypical, or extraordinary costs should be excluded. Some of the items on the above lists could be normal or extraordinary, depending on the facts. For example, most large projects require some overtime labor, which would be a normal expenditure; however, a relatively large amount of overtime to complete a project in a shorter time than normal would be an unusual expenditure that would not be included in the estimate of cost new.

Sources of Data

It is essential that the appraiser have access to an adequate cost and reference library. This library is one of the appraiser's most valuable tools for establishing and supporting an opinion of value. Supporting data are the backbone of a good appraisal, and these data should be available for review by the client or other interested parties.

The MTS appraiser is fortunate to have a number of pricing and reference materials readily available (e.g., manufacturers' and dealers' catalogs, price lists, and equipment specifications). Catalogs containing specifications are an important tool for the appraiser; they can be used to compare one manufacturer's machine to another's and to assist in the determination of cost.

The appraiser's library should be kept current. The organization, cataloging, and updating of the pricing and reference material is a matter of individual preference. Most such libraries are arranged alphabetically by manufacturer or by standard industry code.

A good cost and reference library should contain as much relevant information as can be reasonably obtained. It is suggested that it contain at least the following:

Manufacturers' catalogs

Manufacturers' price lists

Manufacturers' specifications manuals

Published indices

Published price guides

Reference books

Serial number reference guides

Vendors' price lists

Cost and reference materials may be obtained from the following sources:

Engineering and architectural firms
Client invoices
Computerized databases
Contractors
Internet sites
Manufacturer's and vendors
New machinery dealers and representatives
Newspapers
Old files
Other appraisers
Personal contacts
Public libraries
Published price guides
Trade shows and exhibitions
Universities

The appraiser should be aware that the price quoted to him by a supplier may differ significantly from the price that may be quoted to an actual or potential buyer; that manufacturers' discounts can vary greatly; and that delivery and shipping costs should be included or added.

Whatever method is used to obtain cost and reference materials—be it verbal, written, or electronic—rapport must be built with individuals and companies. Good rapport with manufacturers and dealers will facilitate access to these data and generate useful industry contacts. These contacts can provide valuable information on economic trends, present market conditions, and anticipated future developments. Personal contact with other appraisers is one of the best sources of cost and reference data.

Examples of the Direct Cost Method

Examples 3 and 4 illustrate the direct cost method of estimating a property's cost new:

Example 3: Cost of a machine currently in production.

The appraiser's notes indicate the following:

- The subject is a MMM Company engine lathe, model A, size 16" × 30" center-to-center.
- The serial number is 1234 (a serial number guide indicates that the machine was built in 1984).
- The current cost new of the lathe at the MMM Company's factory is $50,500.
- The subject includes an XY taper attachment; the current cost of the taper attachment at the MMM Company's factory is $1,500.
- The motor is 10/1800 with controls (the cost of the motor with controls is included in the basic cost of the lathe).

- The current cost of freight to the subject site is estimated to be $1,800.
- The current cost of installation is estimated to be $3,200.
- The amount of sales taxes is estimated to be $2,800.
- Depreciation or obsolescence is to be ignored for purposes of this example.

The cost new installed of the MMM Company Model A engine lathe can be estimated as shown in table 3.2.

Current Cost of MMM Company Model A engine lathe	$50,500
Add current cost of taper attachment	$1,500
Total Current Cost New at the Manuracturer' Factory	$52,000
Add installation labor	$3,200
Add freight	$1,800
Add sales taxes	$2,800
Total Current Cost New, Installed	$59,800

Table 3.2. Example 3 Solution

In example 3, the appraiser merely needs to include all the normal direct and indirect costs. However, it is often necessary to estimate the cost of an item by assembling many costs or pieces of information. Example 4 illustrates some of the detail that can be required to develop such an estimate.

Example 4: Cost new of a conveyor at the conveyor manufacturer's plant.

The appraiser's notes indicate the following:

Conveyor Components

21'6" long center-to-center, 24" wide

6" formed 308 stainless steel frame

2" diameter by 24" long plastic rollers 12.5" on center.

Roller bed: 2" diameter × 24" long plastic roller return idlers 48" on center

10" diameter by 24" long rubber lagged head drum

6" diameter × 24" galvanized tail drum

24" wide 6-ply sanitary rubber belting

8" diameter sprocket (one required)

4" diameter sprocket with set collar (one required)

1" roller chain drive (36" long)

Angle gear motor drive, 1 horsepower, 48 RPM (revolutions per minute) 220/440-volt alternating current

Six pair 2" square pipe galvanized saddle legs, 30" high.

Wiring Components

One SealTight flexible connection

12 linear feet 1/2" conduit with 3 strands 14-gauge wire

Safety switch, 30 amp, 240 volt.

Other Costs

Labor: The time to fabricate the conveyor at the manufacturer's plant is 2 men for 8 hours each.

Engineering: The time to engineer and design the conveyor is 1 man for 16 hours.

Overhead and profit: All of the manufacturer's overhead and profit is included in the unit costs set out herein.

Costs

- 6" formed 308 stainless steel frame, $15.40 per linear foot
- 2" diameter × 24" long rollers, $24.15 each
- 2" wide 6-ply sanitary rubber belting, $6.95 per linear foot
- 8" sprocket with set collar, $15.35 each
- 4" sprocket with set collar, $9.95 each
- 10" diameter × 24" rubber lagged head drum with pillow blocks and bearings, $69.30 each
- 6" diameter galvanized tail drum assembly complete, $26.75 each
- 1" roller chain, $2.63 per linear foot
- Motor, 1 horsepower, 48 RPM, $765.50 each
- 1 pair 2" square pipe saddle legs, adjustable height, $15.40 each
- 1 SealTight flexible connection, $22.39 each
- 1/2" conduit and wire, 3 strands 14-gauge wire, $5.17 per linear foot
- 30-amp, 240-volt safety switch, $197.65 each
- Labor for fabrication (including all payroll burdens), $33.68 per hour
- Engineering (including all payroll burdens), $38.35 per hour

The cost new to design and fabricate the conveyor can be estimated as shown in table 3.3.

DESCRIPTION		COST
Stainless Steel Frame	43' @ $15.40 LF (2 sides x 2'-6" = 43')	$662.20
Plastic Rollers	20 @ $24.15 Each (21'-6" minus 5" head and 3" tail = 20'-10"÷ 12.5" center-to-center = 20)	$483.00
Return Rollers	5 @ $24.15 Each (21'-6" minus 5" head and 3" tail = 20'-10" ÷ 50" center-to-center = 5)	$120.75
Belting	45' @ $6.95 LF (see calculations in footnote)*	$312.75
Head Drum	1 @ $69.30 Each	$69.30
Tail Drum	1 @ $26.75 Each	$26.75
8" Diameter Sprocket	1 @ $15.35 Each	$15.35
4" Sprocket with Set Collar	1 @ $9.55	$9.55
1" Roller Chain	3' @ $2.63 LF	$7.89
Gear Motor	1 @ $765.50 Each	$765.50
6 Pair Legs	6 pair @ $15.40 Each	$92.40
1 Sealtight Flexible Connection	1 @ $22.39	$22.39
½" Conduit and Wire	12' @ $ 5.17 LF	$62.04
Safety Switch	1 @ $197.65 Each	$197.65
Labor	16 Hrs. @ $33.68 Per Hour	$538.88
Total Direct Cost		
Engineering	16 Hrs. @ $38.35 Per Hour	$613.60
Total Cost or Total Reproduction or Replacement Cost New (At the conveyor manufacturer's plant. Freight & handling to a user not included).		$4,000.00

Table 3.3. Example 4 Solution

** Note: The calculation of the belting in this example requires the estimation of the belting for not only the active part of the conveyor but also the return belt and the portion around the end drums.*

Circumference $= \pi \times D$

Circumference of head $= \pi\,(3.14) \times 10'' = 31.40''$

and then multiply by 0.5 for only ½ of drum $= 15.70''$

Circumference of tail $= \pi\,(3.14) \times 6'' = 18.84''$

and then multiply by 0.5 for only ½ of drum $= \underline{9.42''}$

Total		= 25.12" (Say 2')
	Then 2' + 43' (2 × 21.5' = 43')	
Say		= 45'

Example 4 uses the same techniques used in example 3. The only difference is that the level of detail is much greater in example 4 than in example 3.

In both cases, it could be said that the result is a "like-for-like" cost, that is, reproduction cost new. On the other hand, since the examples do not assume that the hypothetical replacement property would have any design or operating cost advantages, the result could also be described as an "equivalent utility" cost, that is, replacement cost new.

Strengths and Weaknesses of the Detail Method

In theory, the detailed itemizing of cost components should result in greater accuracy than less detailed methods. Another strength of the detail method is that it identifies the cost component having the greatest effect on the total cost of the property. Thus, the direct method is sometimes preferred over methods based on gross costing by unit (sometimes called "unit costing"), for example, cost per linear or square foot.

On the other hand, a problem with the detail method is that it is not always practical, or possible, to create a complete list of either the components or their costs. It is easy to inadvertently omit components that are material to the valuation. Other problems with the detail method:

- Creating the detail is often time consuming and can significantly increase the cost of the appraisal.
- A large amount of detail may not be required either by the client or to accomplish the purpose of the appraisal.
- It may not be possible to itemize all the costs because the required engineering documents may not be available or because the properties are bought and sold as systems or packages.
- The appraiser will learn through experience when to use, and not use, the detail method. If the detail method is

used, the appraiser should review the results to see whether they make sense, such as by comparing them with results produced by other methods.

Trending Method

Trending is a method of estimating a property's reproduction cost new (not replacement cost new) in which an *index* or *trend factor* is applied to the property's *historical cost* to convert the known cost into an indication of current cost. Simply put, trending reflects the movement of price over time.

As used in this book, *historical cost* is the cost of a property when it was first placed into service by its *first owner*. This is to be distinguished from *original cost,* which is the actual cost of a property when acquired by its *present owner,* who may not be the first owner and who may have purchased at a price greater or less than the historical cost.[9] Original cost may be the used cost of the property, whereas historical cost can never be a used cost. Obviously, historical cost and original cost may be the same.

When trending, an index or trend factor is applied to the property's historical cost, not original cost. An *index* is a "number used to measure change in prices, wages, employment, production, etc.; it shows a percentage variation from an arbitrary standard, usually 100, representing the status at some earlier time...."[10] The indices commonly used by appraisers are sets of numbers measuring changes over time in the cost of assets, material, or labor. A sample index is shown in table 3.4.

Year	Index
1995	100
1996	103
1997	105
1998	108
1999	110

Table 3.4. Sample Index

To estimate the current reproduction cost new in 1999 of a machine that had a historical cost in 1995 of $1,000, the calculation would be:

$$\text{Current Reproduction Cost New} = \left(\frac{\text{Current Index}}{\text{Base Year Index}}\right) \times \text{Historical Cost} =$$

$$\text{Current Reproduction Cost New} = \left(\frac{110}{100}\right) \times \$1{,}000 =$$

$$\text{Current Reproduction Cost New} = 1.1 \times \$1{,}000 = \$1{,}100$$

The use of indices and trending can easily lead to erroneous results.[11] The following cautions are offered:

- Indices generally do not reflect technological advances that may actually reduce cost new; thus, trending does not indicate replacement cost new or provide a means to measure the difference between reproduction cost new and replacement cost new.
- Trending should not be applied to anything other than a historical cost, that is, the cost of a property when it was first placed into service by its first owner.
- The appraiser should be certain that the cost to which a trend factor is applied is actually the historical cost, and not a cost resulting from a prior allocation of purchase price or a used cost.
- The historical cost that one is attempting to trend may not be typical cost; that is, it may include extraordinary costs or exclude indirect costs, and so on.[12]
- Cost indices are based on average values, but specific cases (including the subject property) may differ from the average.[13]
- Trending used costs is improper because costs of used equipment are generally not impacted over time at the same rate as new costs.
- The appraiser should be extremely cautious about trending for periods in excess of ten years, unless the accuracy of the trended result can be independently confirmed by comparing it with the results of other methods of

estimating cost new.[14] This is especially true when using a broad-based index or in periods of relatively high inflation.

- The index factor should match the subject property; hence, the user should know at least the basics of how the index was developed.

Cost-to-Capacity Method

A third method of estimating cost new is commonly referred to as *cost-to-capacity* method. According to Jelen and Black:

> The costs of similar plants or pieces of equipment of different sizes vary with the size raised to some power. This relationship can be expressed mathematically as follows:
>
> $$\left[\left(\frac{C_2}{C_1}\right) = \left(\frac{Q_2}{Q_1}\right)^x\right]$$
>
> Here, C_2 is the desired cost of capacity Q_2 and C_1 is the known cost of Capacity Q_1. A frequent value for X is 0.6, so this relationship is often referred to as the *six-tenths (0.6) factor rule....* The exponent X can be determined by plotting actual historical costs for the equipment or plant as the ordinate on log-log paper and the equipment or plant size as the abscissa. The slope of the resulting line through the data will be the cost-capacity factor.[15]

This methodology assumes that not all costs vary with size in a straight line. For example, if the cost of a 10,000-gallon tank is $7,000, it does not necessarily mean that the cost of one twice as big is $14,000. The cost-to-capacity method can be applied to many different items of equipment or to entire facilities. Cost-to-capacity analyses have been prepared for entire process facilities as well as components of facilities, including pumps, tanks, blenders, compressors, heat exchangers, and others. Studies indicate that although cost capacity factors average about 0.6 to 0.7, they vary depending on the property in question. The appraiser should not assume, because of the term *six-tenths factor rule,* that the appropriate factor is necessarily close to 0.6. The factor can be obtained from published data if the published data matches the subject property.[16] If a published exponent is not available, the appraiser can develop his or her own exponent if he or she knows the price and capacity of two or more items similar to the subject.

Examples 5 and 6 show how to use the cost-to-capacity method to estimate the cost new of an entire process plant.

Example 5: Cost-to-capacity when the exponent is known.

Assume it is known that the cost new of an ethylene plant with a capacity of 100,000 tons per year is $16,000,000. Estimate the cost of a 200,000 tons per year plant.

Solution — Using a 0.6 factor (for illustration purposes only):

The formula is:

$$C_2 = C_1\left(\frac{Q_2}{Q_1}\right)^x$$

Where: C_2 = desired cost of capacity Q_2
C_1 = Known cost of capacity Q_1
X = Cost-capacity factor

Solution:

$$C_2 = \$16{,}000{,}000\left(\frac{200{,}000}{100{,}000}\right)^{0.6} = \$24{,}251{,}465$$

Further information on the cost-to-capacity method can be found in engineering textbooks such as *Cost and Optimization Engineering,*[17] *Cost Engineering in the Process Industries,*[18] and *Basic Cost Engineering.*[19]

Example 6: Cost-to-capacity when the exponent is not known.

Calculate the exponent "*x*" in the cost-to-capacity relationship for a 4,000-unit-capacity item, given the information in table 3.5 for two other known units A and B.

Variable	Unit A	Unit B
Size	5,000 = Q2	3,000 = Q1
Price	$171,000 = C2	$119,000 = C1

Table 3.5. Unit Costs

Solution:

$$\frac{C_2}{C_1} = \left(\frac{Q_2}{Q_1}\right)^x$$

$$\frac{\$171{,}000}{\$119{,}000} = \left(\frac{5{,}000}{3{,}000}\right)^x =$$

$1.43697 = 1.66667^x$
$x = 0.7097$ (rounded to 0.71)

With an exponent of 0.71, the cost of a 4,000-capacity unit (C_3) is estimated as follows:

$$C_3 = \$171{,}000\left(\frac{4{,}000}{5{,}000}\right)^{0.71}$$

$$C_3 = \$171{,}000(0.8)^{0.71}$$

$$C_3 = \$171{,}000 \times 0.85348$$

$$C_3 = \$145{,}945$$

Problems with Cost-to-Capacity

The cost-to-capacity method must be used with caution. In fact, many of the cautions expressed above with respect to trending are also applicable to this method. The appraiser should not use the cost-to-capacity formula to scale up a unit or plant beyond what is physically possible. When large differences in capacity are being calculated, significant error can occur if such large differences are misapplied.

Another problem with the method is attempting to extrapolate the results to sizes or capacities that are significantly larger or smaller than the range of the known data. For example, the exponent 0.69 may be valid for squirrel cage induction motors ranging between 5 to 20 horsepower; however, this exponent may not apply to motors of the same kind ranging between 100 to 200 horsepower. For example, if the exponent is 0.6, the error introduced by different capacity ratios is indicated by Jelen and Black to be as presented in table 3.6.

	Actual Cost to Capacity Factor								
	0.2	0.3	0.4	0.5	0.6	0.7	0.8	0.9	1.0
Ratio of Capacity Q2:Q1	Percent Error								
5:1	+89	+61	+37	+17	0	-16	-28	-39	-48
10:1	+150	+100	+59	+26	0	-21	-37	-50	-60

Table 3.6. Percent Error from Improper Use of Cost-to-Capacity

Table 3.6 demonstrates the magnitude of the error from misapplication of the cost-to-capacity method. As the ratios of capacity Q2 to capacity Q1 increase from 5:1 to 10:1, so does the percentage of error. If the appraiser tried to use the exponential factors for other, larger quantity ratios such as 20:1, 50:1, or 100:1, the results could be meaningless.

Other Engineering Methods

Several other engineering methods may be used to estimate the cost of entire facilities or components of facilities; most of these methods are best used in chemical or petrochemical processing industries.

The *Lang factor* method estimates the total facility cost as a function of the equipment cost. For example, if the appropriate factor or multiplier for a particular kind of process facility is 5.5 and the cost of the equipment is $6,000,000, then the total facility estimated cost would be $33,000,000 ($6,000,000 × 5.5).

The *Hand factor* method uses different multipliers for different types of equipment, which may yield a greater accuracy than the Lang factor method. For example, if the appropriate installation multiplier is 8.0 for pumps and 4.5 for tanks, and if the uninstalled cost is $3,000 for the pumps and $50,000 for the tanks, then the *installed* cost of these assets could be estimated as follows:

Pumps: $3,000 × 8.0	=	$ 24,000
Tanks: $50,000 × 4.5	=	$225,000
Estimated Installed Cost	=	$249,000

The appraiser must use caution when applying the Lang or Hand methods, because the same considerations applicable to exponential sizing may also apply to these methods. Some of the size differences are often taken into consideration in the estimation of the basic equipment costs.

In conclusion, there are several methods available to assist in estimating the cost new of process plants or portions of those plants. Among these are the Chilton method, plant cost by analytical procedures method, Peters-Timmerhaus ratio factors method, compartmentalization method, Miller method, turnover ratios method, and investment cost per unit of capacity method. Descriptions of these methods can be found in various cost engineering books and industry handbooks.

In conclusion there are several methods available for estimating the cost new of a property. It is often necessary to use more than one method to estimate the cost new of a plant or other complex set of assets. For example, assume the appraiser knows the cost of a five-year-old plant that is similar in most respects to the subject but of a different capacity; moreover, the subject plant also has a separate division making a product not made at the five-year-old plant. Here the appraiser may use the cost-to-capacity method to establish the plant size relationship, trending to reflect the current cost of the five-year-old plant, and the detail method to add the components of the separate division.

Depreciation from All Causes

Concept of Appraisal Depreciation

In chapter 1, the point was made that the accounting concept of depreciation differs entirely from the appraisal concept of depreciation. Grant defines "appraisal depreciation" as "the difference in value between an existing old property and a hypothetical new property taken as a standard of comparison."[20] He goes on to explain that "appraisal depreciation should measure value inferiority" and that

> The most economical new substitute property may have many advantages over an existing old property, such as longer life expectancy, lower annual disbursements for operation and maintenance, increased receipts from sale of product or service. The depreciation deduction from the cost of the hypothetical new substitute property should be a measure in money terms of all of these disadvantages of the existing old property.[21]

Types or Causes of Depreciation

The three types or causes of appraisal depreciation traditionally recognized by appraisers are *physical deterioration, functional obsolescence,* and *economic obsolescence.* The traditional definitions of these terms are as follows:

> *Physical deterioration* is the loss in value or usefulness of a property due to the using up or expiration of its useful life caused by wear and tear, deterioration, exposure to various elements, physical stresses, and similar factors.
>
> *Functional obsolescence* is the loss in value or usefulness of a property caused by inefficiencies or inadequacies of the property itself, when compared to a more efficient or less

costly replacement property that new technology has developed. Symptoms suggesting the presence of functional obsolescence are excess operating cost, excess construction (excess capital cost), over-capacity, inadequacy, lack of utility, or similar conditions.

Economic obsolescence (sometimes called "external obsolescence") is the loss in value or usefulness of a property caused by factors external to the property, such as increased cost of raw materials, labor, or utilities (without an offsetting increase in product price); reduced demand for the product; increased competition; environmental or other regulations; inflation or high interest rates; or similar factors.

Physical Deterioration

Once the proper level of current cost new has been determined, deductions must be made for all forms of depreciation. Normally physical deterioration is considered first. Physical deterioration is the loss in value or usefulness of a property due to the using up or expiration of its useful life caused by wear and tear, deterioration, exposure to various elements, physical stresses, the passage of time, and similar factors. It is generally a result of the expiration of the property's useful life over time, exposure to natural elements or the process area environment, internal defects from vibration and operating stress, and similar factors. Excessive deterioration often reduces tolerances on manufacturing equipment, resulting in higher product rejection and ultimately the inability to meet production standards. Dealing with the effects of physical deterioration often requires the owner to make higher-than-average maintenance expenditures. On the other hand, below-average maintenance expenditures may indicate the possibility of deferred maintenance and increased deterioration.

Physical deterioration is often estimated as a percentage: a new property has zero percent physical deterioration, while a property that is completely exhausted and has no scrap value has one hundred percent physical deterioration, with all other properties falling between these two extremes. Theoretically, physical deterioration can be measured objectively. A machine will produce *x* number of parts in its physical life. Assuming that adequate statistics were kept, the machine was never rebuilt or abused, and all assets of the same type are equivalent, then a simple ratio of the past production to the expected production would result in an objective measure of physical deterioration. Obviously, machines are rebuilt, they are used and abused, and their production quality varies. Except for large or high-cost assets, production statistics are usually not available for individual assets. Therefore, measuring physical deterioration is often subjective. Often the appraiser must rely on how similar assets have performed in the past in order to make judgments of the physical condition of the subject.

Measuring Physical Deterioration

The best procedure to follow when measuring physical deterioration is to rely on the facts and circumstances applicable to the subject, particularly the age and use of the property. When trying to measure only one type of accrued depreciation, the appraiser should strive to carefully segment the three elements of depreciation—physical, functional, and economic—and should deal with each element individually. When trying to estimate only physical deterioration, the appraiser should avoid inadvertently including functional or economic obsolescence in his or her measurement of "physical deterioration."

The use of a property is often a good indicator of physical deterioration. Machines used 24 hours per day will physically deteriorate faster than the identical units operated eight hours per day. Machinery employed in dusty, dirty, abrasive, or corrosive atmospheres will deteriorate faster than the same property in a clean environment. Machines that have just been rebuilt are in better physical condition than those that need to be rebuilt. A fresh coat of paint does not change the physical condition of a property. It may suggest a high level of maintenance, but new paint may also be used to cover defects. The appraiser should go beyond casual observation to investigate any extraordinary care and maintenance or abnormal wear and tear.

When a property such as a machine tool is rebuilt, its physical deterioration is partially, but not completely, corrected because the rebuilt machine is not new and some physical deterioration (incurable) exists that cannot be corrected. This difference in value can be measured by comparing the selling price of a new unit with the price of one rebuilt and offered for sale by a used equipment dealer.

In some industries, certain maintenance practices are performed that will aid the appraiser in measuring physical deterioration. In the petrochemical industry, ultrasonic inspections of wall thickness for assets such as vessels, tanks, columns, and reactors are common practice. Tank bottoms and tops are replaced more frequently than tank walls. In smelting operations, the refractories in melting furnaces are replaced regularly, while the shell suffers very little wear. These kinds of maintenance practices can be investigated and used in estimating physical deterioration.

When estimating physical deterioration, the appraiser should strive for consistency. While opinions will differ as to the condition of the same asset, a systematic approach by an appraiser allows others reviewing the work to see the asset from the appraiser's viewpoint.

Three methods of measuring physical deterioration that will be discussed here are *observation, formula/ratio,* and *direct dollar measurement*. The following discussion suggests some ways to objectively measure physical deterioration. Ultimately, however, it must be said that isolating and measuring just physical deterioration is sometimes difficult. Nevertheless, the appraiser must address the issue and should strive to base his or her opinion on the actual facts and circumstances.

Observation

In this method, the appraiser makes a comparison based on the experience he or she has gained by looking at similar properties and comparing them to new properties. The procedure involves actually observing those elements of wear and tear that can be seen and converting those observations into a percentage. It also involves discussions with knowledgeable plant personnel to determine the condition of those things that might not be readily apparent, such as internal corrosion on tanks. On the basis of the facts, the appraiser must develop an opinion of physical deterioration, stated in the form of a percentage, to deduct from replacement or reproduction cost new.

Use/Total Use

Another method of estimating physical deterioration involves analysis based on a property's use. Use is a good indicator of physical deterioration when the requisite production statistics can be obtained. In its simplest form, physical deterioration can be estimated by the ratio of

$$\frac{\text{Use}}{\text{Total Use}}$$

Given some unit of measure, this ratio measures the amount of use of a property at a point in time compared with the total use expected from that asset. For example, assume that the normal physical life of a machine is 100,000 hours. If a specific machine has logged 40,000 hours, it is logical to conclude that under normal conditions, the physical deterioration is approximately 40 percent ([40,000 ÷ 100,000] × 100 = 40 percent).[22]

In the case of a new asset, total use is synonymous with the expected use that in this example is 100,000 hours. Purists will argue, and rightfully so, that considering depreciation in this manner is a linear or straight-line calculation of deterioration. In other words, at 50,000 hours, the machine would be depreciated 50 percent, at 60,000 hours, 60 percent, and so on. This implies that when the machine reaches 100,000 hours it is 100 percent or completely depreciated. A contradiction arises when the machine has 100,000 hours and is still operating. When this occurs, *total use* in the above ratio is defined as the sum of the usage to date (100,000 hours) plus the remaining useful life, measured in hours. If the machine has 100,000 hours and is expected to last another

25,000 hours, the physical deterioration would then be 80 percent, developed as follows:

$$\left(\frac{100{,}000}{100{,}000+25{,}000}\right) \times 100 = \left(\frac{100{,}000}{125{,}000} \times 100\right) = 80 \text{ percent}$$

Now consider the situation where the machine has 75,000 hours, has just been rebuilt, and has an expected remaining life of 50,000 hours. Physical deterioration would then be developed as follows:

$$\text{Physical Deterioration} = \left(\frac{75{,}000}{75{,}000+50{,}000} \times 100\right) = 60 \text{ percent}$$

Age/Life

The ratio of a property's "age" to its "life" can be used to measure physical deterioration. Although this is straight-line depreciation, it should not be confused with accounting depreciation because the appraiser uses valuation rather than accounting concepts of age and life.

> Depending on the purpose of the appraisal analysis, different appraisal definitions of age and life can be used as the numerator and denominator in the "age/life" ratio. The various definitions will be presented at this point to facilitate our discussion, although some of the terms will not be used until later.
>
> *Chronological age* is the number of years that have elapsed since a property was originally built or placed in service.
>
> *Effective age* is the apparent age of a property in comparison with a new property of like kind; that is, the age indicated by the actual condition of a property. In estimating effective age, the appraiser considers the effect that overhauls, rebuilds, and above-average or below-average maintenance may have on the property's current condition. If a property has received regular overhauls, its effective age will normally be less, often significantly less, than its chronological age. Effective age

is often the more appropriate numerator in the age/life ratio than is chronological age.

Normal useful life is the estimated number of years that a new property will actually be used before it is retired from service. A property's normal useful life relates to how long similar properties actually tend to be used, as opposed to the more theoretical economic life calculation of how long a property can profitably be used. The best evidence of normal useful life is statistical or actuarial data derived from the study of properties that are similar to the subject under actual operating conditions. An asset's useful life may be longer than its *economic life* because the owner may elect not to retire the asset from service upon expiration of the asset's theoretical economic life.

Remaining useful life is the estimated period during which a property of a certain effective age is expected to actually be used before it is retired from service. The best evidence of remaining useful life is statistical or actuarial data derived from the study of properties that are similar to the subject under actual operating conditions. Remaining useful life can sometimes be approximated by deducting the asset's *effective age* from its *normal useful life*; however, this is an over-simplification and not technically correct in some situations, for the same reason that a person who had a 72-year life expectancy when born, and is now 70 years old, has more than a 2-year remaining life expectancy according to human mortality tables. Statistical and actuarial studies of asset useful lives indicate that many assets follow a similar pattern.

Physical life is the estimated number of years that a new property will physically endure before it deteriorates or fatigues to an unusable condition purely from physical causes, without considering the possibility of earlier retirement due to functional or economic obsolescence.

Remaining physical life is the estimated period during which a property of a certain effective age is expected to physically endure before it deteriorates or fatigues to an unusable condition purely from physical causes, without considering the possibility of earlier retirement due to functional or economic obsolescence.

Economic life is the estimated number of years that a *new* property may be profitably used for the purpose for which it was intended. Stated another way, economic life is the estimated number of years that a new property can be used before it would pay the owner to replace it with the most economical replacement property that could perform an equivalent service. Functional or economic obsolescence factors may limit a property's economic life. An asset's economic life will often be less than its *normal useful life*.

Remaining economic life is the estimated period during which a property of a certain effective age is expected to continue to be profitably used for the purpose for which it was intended. It can be approximated by deducting the asset's *effective age* from its *economic life*, although this is an over-simplification and not technically correct in some situations for the same reasons given above with respect to *remaining useful life*.

In its simplest form, the age/life formula can be used to estimate physical deterioration using the following formula:

$$\frac{\text{Effective age}}{\text{Physical life}} = \text{\% of physical deterioration}$$

For example, if a property has an effective age of 8 years and a physical life of 20 years, physical deterioration will be 40 percent (but see footnote 22).

For larger, older, or more complex property, the age/life concept can be expanded as follows:

$$\frac{\text{Effective age}}{\text{(Effective age + remaining physical life)}}$$

It is essential that the appraiser understands that some of the definitions given above, if substituted in the age/life equation, will measure more than just physical deterioration. Measuring more than physical deterioration is often appropriate, as long as the appraiser is aware of what he or she is doing.

Effective age has been defined as the age indicated by the actual condition of a property. In estimating effective age, the appraiser considers the effect that overhauls, rebuilds, and above- or below-average maintenance may have on the asset's current condition. If a property has received regular overhauls, its effective age will normally be less, often significantly less, than its chronological age.

An advantage of using the age/life technique is that effective age can often be calculated using the client's fixed asset records. The effective age can be determined by weighting the investment in a property or a group of assets. The weighting should be done on some equitable basis and must consider additions or deletions made over the asset's life. The procedure can be used for a single asset (if the records are sufficiently detailed) or, more commonly, for groups of assets. The next example illustrates a method of estimating effective age.

Example 7: Effective age.

Your task is to determine the effective age of a property being appraised as of January 1997. Your investigation reveals that the property was purchased new in 1987 and that capital expenditures for additions were made in 1990 and 1992. In addition, a major overhaul was completed in 1995 that effectively replaced some of the original equipment.

The first step is to develop the proper basis for comparison, which in this case is trended historical cost. This is determined by applying the appropriate cost index to the historical cost for each year (see table 3.7).

Purchase Date	Historical Cost	times	Index Translator	equals	Trended Historical Cost
1987	$20,000	x	2.60	=	$52,000
1990	$2,000	x	1.95	=	$3,900
1992	$2,500	x	1.60	=	$4,000
1995	$17,500	x	1.20	=	$21,000
Totals	$42,000				$80,900

Table 3.7. Application of Cost Index

The total historical cost of $42,000 and the total trended historical cost of $80,900 shown in table 3.7 are somewhat misleading because they include some redundant investment resulting from the overhaul in 1995. In other words, these total costs include not only the initial costs but also the costs of those assets that were replaced as part of the 1995 overhaul. For example, if the $17,500 expended in 1995 was to replace a pump, the cost is included twice: as part of the original investment in 1987 and again in 1995. To adjust for this duplication, it is necessary to delete the 1987 dollars. This is accomplished by restating the 1995 overhaul in 1987 dollars by "back-trending" as follows:

$$\left(\$17{,}500 \times \frac{1.20}{2.60}\right) = \$8{,}076 \approx \$8{,}100 \text{ rounded}$$

The historical cost of $20,000 in 1987 minus $8,100 equals $11,900, which is the restated historical cost less the redundant investment. The calculations to this point are summarized in table 3.8.

Purchase Date	Adjusted Historical Cost	times	Index Translator	equals	Trended Historical Cost
1987	$11,900	x	2.60	=	$30,940
1990	$2,000	x	1.95	=	$3,900
1992	$2,500	x	1.60	=	$4,000
1995	$17,500	x	1.20	=	$21,000
Totals	$33,900				$59,840

Table 3.8. Trended Adjusted Historical Cost

The next step is to weight the trended historical costs (as adjusted) using their age. This is done by multiplying each trended historical cost by its chronological age (see table 3.9).

Purchase Date	Trended Historical Cost	times	Chronological Age of Investment, Years	equals	Weighted Investment
1987	$30,940	x	10	=	$309,400
1990	$3,900	x	7	=	$27,300
1992	$4,000	x	5	=	$20,000
1995	$21,000	x	2	=	$42,000
Totals	$59,840				$398,700

Table 3.9. Determination of Weighted Investment

The last step is to determine the composite effective age. This is done by dividing the total weighted investment by the total trended historical cost, as follows.

$$\frac{\$398{,}700}{\$59{,}840} = 6.66 \text{ years} = 7 \text{ years (rounded)}$$

The rounded result of 7 years is a reasonable estimate of the effective age of the assets being appraised.

The problem presented in example 7 has been simplified to illustrate techniques and concepts. Cost information was used as an equitable basis for comparison. There are other bases that may be appropriate. For example, when trying to estimate the age of a building, an appraiser may wish to develop the effective age on the basis of building area; for equipment, the appraiser may consider estimating the effective age on a capacity basis.

There are some simpler methods that are not as accurate as the method illustrated above. One of these shortcut techniques is to use the trended historical cost information to develop a composite cost index and interpolate using the cost index. If we did that using the data in table 3.9, the composite cost index derived by dividing the trended historical cost by the historical cost (as adjusted in Table 3.8) is 1.765. Interpolating the 1.765 against the cost indices as of 1.60 and 1.95, or 1990 and 1992, indicates a weighted investment date of approximately 1991, or 6 years (the property is being appraised as of 1997).

Another technique that could be used to estimate effective age is weighting the "adjusted historical cost" from table 3.8 based on age (i.e., adjusted historical cost × age in years). If this method is used with the same facts as above, the effective age is 5.3 years (see table 3.10).

Date	Adjusted Historical Cost	times	Chronological Age of Investment, Years	equals	Weighted Investment
1987	$11,900	×	10	=	$119,000
1990	$2,000	×	7	=	$14,000
1992	$2,500	×	5	=	$12,500
1995	$17,500	×	2	=	$35,000
Total	$33,900				$180,300

Table 3.10. Alternative Determination of Effective Age

$$\frac{\$180,300}{\$33,900} = 5.3185$$

The different result produced by this method, compared to the effective age of 6.66 years calculated in example 7, is caused by the different assumptions underlying the weighting. The method shown in example 7, resulting in an effective age of 6.66 years (which was rounded to 7 years), is the most accurate because it measures the age of the asset on an equitable, trended cost basis. Establishing a composite cost index and interpolating is not as accurate because of the inaccuracy inherent in the interpolation process, as well as the fact that there may be short-term aberrations in the cost index. Another method occasionally seen (and not discussed above), that of using untrended historical cost multiplied by age, is wrong and inaccurate because it gives equal weight to each cost, regardless of the year in which the cost was expended. The degree of error can be substantial when the effective age calculation involves a long span of years or a period during which inflation was significant.

The advantage of the ratio technique demonstrated in the example is that it is possible to reasonably estimate the effective age of a property, assuming that the required data are accurate and available. Even using this method, the problem that remains is estimating the remaining physical life of that asset. If some physical defects are known that will limit the physical life, a reasonable estimate can be made.

Example 8: Determination of physical condition.

You are appraising a process furnace that operates continuously 24 hours a day, seven days a week, and you are trying to estimate its physical condition. In discussions with the operating and maintenance personnel, you have determined the following: that the furnace has operated normally since it was installed approximately 12 years ago; that some patching on the flues and ductwork was done approximately five years ago; that some pumps, piping, and other external equipment were replaced approximately two years ago; and that the refractory components will need to be replaced in approximately five years.

For this example, assume that you calculated the effective age using the method shown in example 7 and you found it to be eight years. Further, assume that the cost of the furnace components breaks down to 30 percent for the refractory, 50 percent for the structural members, and 20 percent for the other equipment.

The next step would be to estimate the composite physical remaining life for the entire furnace. You have learned that once the refractory is replaced (in five years), the furnace can readily operate for another 15 years, so the physical remaining life of the structural members is 20 years (15 + 5 years). Although the other equipment is in relatively good condition, the plant personnel do not expect it to last as long as the structural members will last. They expect to perform some maintenance on this other equipment to extend its life. The composite physical remaining life for the entire furnace can be developed as shown in table 3.11:

	Percent of Investment	times	Estimated Physical Remaining Life	equals	Weighted Remaining Life
Refractory	30%	X	5	=	1.5
Structural Members	50%	X	20	=	10.0
Other Equipment	20%	X	15	=	3.0
Totals	100%				14.5

Table 3.11. Composite Physical Remaining Life

The composite physical deterioration is simply the ratio of the effective age (8 years) divided by a denominator that is the sum of the effective age and the remaining physical life (8 + 14.5 years), which computes to 35.5 percent as follows:

$$\text{Percent Physical Deterioration} = \frac{8}{8 + 14.5} = \frac{8}{22.5} = 35.5\%$$

Note that this is an estimate of the physical condition of the furnace as it exists. It does not consider the cost of the additional work that would be necessary to extend the life beyond that which is currently anticipated.

Using Age/Life to Estimate Total Depreciation (Physical, Functional, and Economic)

The age/life technique can also be used to estimate more than just physical deterioration, when supported by the facts. For example, assume that you are appraising a pollution abatement facility and a law has been passed requiring the owner to abandon the facility within three years and replace it with a different facility. Furthermore, assume the existing assets are five years old and have a physical remaining life of 15 years. In this case physical deterioration would be 25 percent ($5 \div 20 \times 100 = 25$ percent). Since this asset has a three-year remaining economic life because of the environmental mandate, the total (overall) depreciation (physical, functional, and economic) is approximately 62.5 percent, developed as follows:

$$\text{Total Depreciation Percent} = \frac{5}{5 + 3} = 0.625 \times 100 = 62.5\% = 63\% \text{ (rounded)}$$

Another example of calculating total depreciation appears later in example 13. Again, it is necessary to use this concept in light of the facts and circumstances that affect the property being appraised.

The age/life technique discussed here is based on straight-line valuation depreciation concepts (not straight-line accounting depreciation). The analysis establishes the position of the subject property at an appropriate point in its life cycle. When adjustments are made for observed conditions relating to past and future use, this analysis becomes a valid tool in the appraisal process. It should be noted that the age/life method is a relatively simple way to look at age, life, and use of a property. When sufficient data are available, there are more sophisticated quantitative methods for measuring some kinds of depreciation. For additional information, the MTS appraiser is directed to *Engineering Valuation and Depreciation,*[23] which discusses additional methods of measuring value loss caused by expiration of service life and estimating asset lives, including the Iowa-type survivor curves.

Direct Dollar Measurement

The age/life technique is most useful for newer properties or properties in midlife. When a property requires a significant expenditure to solve a physical problem or a particular component has a short remaining physical life, the appraiser can use this information to estimate physical deterioration by the *direct dollar measurement* method. This concept involves measurement of the dollar amount of physical deterioration. It is applicable when specific components have deteriorated and can economically be cured. This kind of deterioration is *curable.* Deterioration or depreciation is curable when it is economically feasible to remedy it, because the resulting increase in utility and value is greater than the cost to cure. Deterioration or depreciation is *incurable* when it is *not* economically feasible to remedy it, because the resulting increase in utility and value is less than the cost to cure. Examples of deterioration that is usually curable are replacement of a motor, sandblasting and painting of a tank, or rebuilding of a machine tool. An example of incurable deterioration is when metal fatigue has affected a property.

When using this concept, the appraiser should strive to segregate the *curable* elements from the *incurable* elements. The primary difference between the two is that the curable physical deterioration can economically be cured while the incurable portion cannot. By segregating these elements, we are analyzing the property in two parts. The advantage is that the portion that is curable can usually be estimated directly in dollars, providing yet another clue in estimating physical deterioration. Those elements that are incurable must be depreciated based on observation, "age/life," or "use/total use" techniques described earlier. The sum of the curable and the incurable deterioration represents the total physical deterioration of the subject property.

Example 9: Curable and incurable physical deterioration.

Your task is to appraise a crude oil tank that is part of a tank farm in an oil refinery, under a market value in continued use concept using the cost approach. You have determined that the storage capacity requirements of a modern refinery are the same as those of the subject. The only difference is that a modern refinery would use fewer, but larger, individual tanks to meet those storage requirements. A replacement cost new is developed based on the configuration of the storage requirements of the modern refinery. You have concluded that the replacement cost new is approximately 10 percent less than the reproduction cost new for the equivalent storage capacity. For this example, then, the replacement cost new for the individual tanks is simply the reproduction cost new minus 10 percent.

During your discussion with the maintenance personnel, you learned that one crude oil tank has a small leak in its bottom because of corrosion from the saltwater that is naturally present in this type of crude oil. The tank has been pumped out and is being prepared for maintenance. Preliminary indications are that the entire bottom as well as the first course of the side walls will need to be either patched or replaced, depending on the wall thickness. The planned expenditure for this repair work is approximately $350,000, which includes the costs associated with removing the tank from service, cleaning, preparing a safe work environment to make the repairs, and removal and replacement of the corroded areas.

You have determined that the reproduction cost new for this tank is $2,000,000; since replacement cost new is 10 percent less than reproduction cost new, its replacement cost new is $1,800,000.

The next step is to determine the physical deterioration. In this case, we have a physical problem that is curable: replacement or patching of the tank bottom and portions of the side walls. The total physical deterioration of the tank should be deducted from reproduction cost new, since this is the specific property that is being appraised. The result, expressed as a percentage of reproduction cost new, should be applied to replacement cost new. In this way, we are measuring the actual physical deterioration and deducting it proportionally from replacement cost.

The first step is to deduct the cost to cure the physical problem (see table 3.12).

Reproduction Cost New	$2,000,000
Less Curable Physical Deterioration	–$350,000
Reproduction Cost New of Incurable Portion of Tank	$1,650,000

Table 3.12. Curable Physical Deterioration

The $1,650,000 remaining is the cost of the portion of the tank subject to incurable deterioration. The incurable deterioration may be estimated using the observation or formula/ratio techniques described earlier.

Assume that the tank is ten years old and that we estimate at least another 15 years of remaining physical life (remember that these areas of the tank would not be affected by the saltwater). The total incurable physical deterioration is 40 percent, or \$660,000, calculated as follows:

$$\left(\frac{10\text{ Years Effective Age}}{10\text{ Years Effective Age} + 15\text{ Years Remaining Useful Life}}\right) \times 100 =$$

$$\left(\frac{10}{25}\right) \times 100 =$$

$$(0.40 \times 100) = 40\text{ percent}$$

The incurable physical deterioration is then applied to the incurable portions as follows:

Reproduction Cost New Incurable Portion	\$1,650,000
Incurable Physical Deterioration Percentage	× 40.0%
Incurable Physical Deterioration	\$660,000

Table 3.13. Incurable Physical Deterioration

The sum of the curable physical deterioration of \$350,000 and the incurable physical deterioration of \$660,000 is \$1,010,000, which when divided by the reproduction cost new suggests a composite physical deterioration of 51 percent, as follows:

$$\text{Composite Physical Deterioration} =$$

$$\left(\frac{\text{Curable} + \text{Incurable Deterioration}}{\text{Reproduction Cost New}}\right) \times 100 =$$

$$\left(\frac{\$350{,}000 + \$660{,}000}{\$2{,}000{,}000}\right) \times 100 =$$

$$0.505 \times 100 = 51\text{ percent (rounded)}$$

The composite physical deterioration of 51 percent should then be applied against the replacement cost new (see table 3.14).

Reproduction Cost New	$2,000,000
Less 10%	–$200,000
Replacement Cost New	$1,800,000
Less Physical Deterioration at 51%	–$918,000
Cost New Less Physical Deterioration	$882,000

Table 3.14. Cost New Less Composite Physical Deterioration

We have discussed several methods of estimating physical deterioration: observation, age/life, use/total use, and direct dollar. All are valid given certain facts, sufficient information, and an appropriate amount of time in which to analyze the data. In theory, all the methods should be considered, but this is not always practical. The facts and circumstances will dictate the appropriate method or methods to be used in the appraisal assignment.

Sources of Information for Physical Deterioration

The primary sources of information for determining physical deterioration include historical and future capital expenditures; production records; discussions with maintenance and engineering personnel; and outside sources such as manufacturers, dealers, and service representatives. By reviewing the historical capital expenditures, the appraiser may see that substantial money has been spent recently on a particular asset, suggesting that it is now in reasonably good condition. Perhaps very little money has been spent, which may be normal for that asset or may suggest that the asset needs major maintenance. When analyzing a company's budget for future expenditures, an appraiser can readily see where major expenditures are anticipated. These expenditures may offer clues as to which properties are approaching the end of their useful physical lives or suffering from functional or economic obsolescence. A review of production records will provide an indication of how particular properties have been used. If production is down, it may be because of some physical problem. If production is up for an extended period, this may suggest an increase in physical deterioration that will either shorten the life of the property or require major maintenance.

There are no hard and fast rules regarding what should be ascertained when reviewing records. Records should provide some clues and ultimately lead to the facts on which to base an opinion. After the records have been reviewed, the appraiser should spend time discussing conditions with various maintenance and engineering personnel. It is difficult for an appraiser, or any outsider, to ascertain the condition of a facility without talking to those individuals who operate and maintain the assets on a daily basis. These conversations will help the appraiser to assemble additional information and facts upon which to base an opinion.

A word of caution is necessary regarding future capital expenditures: when reviewing documents like five-year plans, it is quite common to see projections for the total replacement of certain properties. Often these replacements are made for reasons other than physical deterioration, and the appraiser should strive to ascertain the reasons why those particular properties are being replaced. Some of the future capital expenditures may be for repair and maintenance of the existing property, and some may be earmarked for expansion. If a complete replacement of a property or group of assets is planned in the future, it may relate to the asset's physical condition, or the replacement may be needed to solve some functional or economic problem. If the latter is the case, then the asset scheduled for replacement may be affected by functional or economic obsolescence. We will now turn to functional obsolescence.

Functional Obsolescence

The next step in implementing the cost approach is to consider *functional obsolescence*. Functional obsolescence has been previously defined as the loss in value or usefulness of a property caused by inefficiencies or inadequacies of the property itself, when compared to a more efficient or less costly replacement property that new technology has developed. Symptoms suggesting the presence of functional obsolescence are excess operating (i.e., manufacturing) cost, excess construction (excess capital cost), over-capacity, inadequacy, lack of utility, or similar conditions.

Some appraisers draw a distinction between functional obsolescence and *technological obsolescence*. They define functional obsolescence as a loss in value resulting from differences in capability between a new machine and the appraised machine, and technological obsolescence as a loss in value resulting from the difference between design and materials of construction used in present-day machines compared with those used in the machine being appraised. There is a legitimate difference of opinion as to how appraisers apply the concepts to measure the functional and technological aspects affecting value. Regardless of the terms used, the important thing is for the appraiser to measure the various factors that contribute to functional obsolescence.

Functional Obsolescence from Excess Capital Costs

Functional obsolescence resulting from *excess capital costs* has already been discussed. Assuming that a property's replacement cost is less than its reproduction cost, which is often but not always the case,[24] excess capital cost is measured by the difference between reproduction and replacement cost. This excess cost represents the decreased capital investment required to obtain the most economical new property to perform the same service as the subject. Functional obsolescence due to excess capital costs results from improvements and changes in design, materials, layout, product flow, construction methods, and equipment size and mix. Essentially, these are the improvements that make the new technology more desirable.

Functional Obsolescence from Excess Operating Costs[25]

The second type of functional obsolescence is that caused by excess operating or manufacturing costs, which will generally be referred to as *operating obsolescence*. As a result of new technology, in many cases not only is it cheaper to acquire a modern replacement property (capital cost), but it is also cheaper or more efficient to operate the new property (operating cost). Calculating operating obsolescence involves a comparison of the operating characteristics of the subject property to its modern equivalent, the property with which the subject would (hypothetically) be replaced. The existing property or facility and its higher operating costs are compared to the reduced costs that could be achieved by the hypothetical replacement facility. From this comparison, a basis for estimating a penalty for continued use of the existing property is developed.

An operating obsolescence study involves the following steps:

1. Analyze the operating statements of the subject property to determine its operating cost per unit of production.

2. Determine the operating costs per unit for the modern replacement facility.

3. Determine the difference in operating cost per unit.

4. Convert the operating cost differential into the total annual excess operating cost that will be incurred by the subject as a result of continued operation. This may be accomplished by applying the operating cost differentials to projected annual capacities or by some other method consistent with the facts and circumstances.

5. Reduce the total annual excess operating cost to reflect the effect of income taxes on the incremental income (i.e., the additional operating income that the modern replacement plant would achieve due to its lower operating cost).

6. Estimate the remaining life during which the subject will continue to be penalized by the excess operating cost (i.e., the time during which the excess operating costs will continue to exist).

7. Convert the annual excess operating cost to present value using an appropriate discount rate that reflects the risk of the excess operating costs that are being discounted.[26]

Operating obsolescence is defined as the present value of the future excess operating costs. The subject property will continue to incur excess operating costs, over and above those of a modern facility, until the problem is corrected, the assets wear out, or, in extreme cases, the company goes out of business. Essentially, the appraiser is comparing the operating efficiency of an existing plant or asset to its modern equivalent, the modern asset that would replace it. The replacement property should be the most economical asset that would replace the service provided by the subject property.

Operating obsolescence is independent of any physical deterioration or functional obsolescence due to excess capital costs that may affect the subject property. Some appraisers incorrectly assume that operating obsolescence double counts the functional penalties. The logic is that the appraiser has measured all functional obsolescence by starting with replacement cost new instead of reproduction cost new. This is not correct. Starting from replacement cost new eliminates functional obsolescence caused by excess capital costs, but it does not take into account the future lost profits the buyer will incur after buying the subject with its excess operating costs.

Operating obsolescence is calculated on an after-tax basis, based on the assumption that the excess operating cost incurred by the subject would be taxable income in the modern plant (i.e., the modern plant would realize the excess operating costs as additional operating profit, which would be taxable).

The kind of operating costs that should be investigated for the existence of operating obsolescence are operating labor, maintenance labor and materials, operating supplies and chemicals, raw material cost, energy and utilities, and production yields.

Functional obsolescence, particularly operating obsolescence, is typically found in the following situations:

1. Plants involved in the process industries

2. Plants involved in industries that either use assets or manufacture products with a high degree of technology

3. Older plants that have increased in size over time

4. Plants in which there are a number of identical units

5. Plants involved in industries which handle large volumes of material

6. Plants with areas of inactive machinery

Example 10: Measurement of functional obsolescence in example 1.

Earlier in this chapter, example 1 discussed how to determine the replacement cost new for four boilers that were part of a petrochemical refinery. Now we consider the operation of those boilers and look at the operating obsolescence (i.e., the functional obsolescence from excess operating costs) associated with them.

In addition to the information obtained in example 1, you learn that 19 employees are required to operate the existing boiler facilities and that these individuals are assigned to each boiler as indicated in table 3.15.

Boiler Number	Capacity (pounds per hour)	Number of Employees*
1	15,000	0
2	30,000	4
3	30,000	7
4	60,000	8
Total	135,000	19

* *Note: Average over a one-year period.*

Table 3.15. Employees Assigned to Boilers

From your interview with the manager of the modern refinery that your client had recently built in a neighboring state (see facts of example 1), you learn that it requires only ten employees (9 less than in the subject facility) to operate the boiler facilities in the modern refinery. (Note that in this example, as well as in practice, the modern equivalent is more desirable because it is cheaper to build and, from a labor standpoint, cheaper to operate.) The nine extra people required in the subject facility represent additional production costs for the subject property compared with its modern counterpart. This additional cost places the subject at a disadvantage. This disadvantage, the annual excess operating costs, can be quantified by multiplying the number of excess employees times the total labor cost per employee. The plant controller tells you that the direct labor rate for each excess employee, as of the appraisal date, averaged $25,000 per year; and that the benefits paid by the company (insurance, FICA [Federal Insurance Contributions Act], and other benefits) amount to 30 percent of the direct labor rate, or $7,500. Therefore, the total cost per person is $32,500 per year, which equals $292,500 per year for the nine people. This represents the annual excess operating cost (pre-tax) associated with the subject facility.

The next step is to convert the annual excess operating cost to present value. For this example, assume that the boilers have a remaining useful life of ten years, the combined federal and state income tax rate is 40 percent, and the discount rate is 10 percent.

The operating obsolescence is estimated as shown in table 3.16.

Annual Excess Operating Costs	$292,500
Less Taxes at 40%	$117,000
Annual Excess Operating Cost After Tax	$175,500
Present Value Factor of 10% for 10 Years	6.14457
Operating Obsolescence (Labor)	$1,078,372
Rounded to	$1,078,000

Table 3.16. Estimation of Operating Obsolescence

This is a relatively simple example. In practice, properly measuring operating cost differentials requires experience and a significant amount of research. The point is that if an individual asset, or an entire plant, produces products at a greater cost than the asset that would replace it, then the existing asset is less desirable and hence less valuable. Analysis of the operating obsolescence allows the appraiser to quantify the value inferiority of the subject, when compared to its modern replacement.

Example 11: Operating obsolescence.

Example 2 illustrated a situation in which replacement cost new was greater, instead of lower, than reproduction cost new. The primary reason for this in example 2 was that an advancement in technology had resulted in a replacement machine that would produce each unit at a lower operating cost, although the replacement machine's capital cost (i.e., the subject's replacement cost new) was greater than the subject machine's reproduction cost new. In that example, the capacity of the subject (Model A) and its modern replacement (Model B) were both 100 units per day. However, Model A required 100,000 British thermal units (BTUs) per year, whereas Model B required only 75,000 BTUs per year, a difference of 25,000 BTUs. At 50 cents per BTU, Model B generates an annual operating cost savings of $12,500 per year; or stated another way, Model A incurs excess operating costs of $12,500 per year.

Assume you have analyzed the physical condition of the subject (Model A) and have estimated its effective age as six years. Your research suggests that the subject's normal physical life is approximately 15 years. On the basis of the information you have obtained, there is no reason to conclude that the normal useful life will be shorter or greater than 15 years, which implies a remaining physical life of nine years (15 minus 6 = 9 years). Therefore, physical deterioration is 40 percent (6 ÷ [6 + 9] = 40 percent).

The next step is to determine the period of time during which the excess operating cost penalty of $12,500 per year will continue to be incurred by the subject. Discussions with the operating personnel suggest that a nine-year physical remaining life is a reasonable estimate. The operating personnel agree that at the end of the subject's physical life, major expenditures will be required to improve its physical condition as well as its operating performance.

The operating personnel also agree that because of the advancement of technology, the subject property is less desirable from an operating standpoint. Nevertheless, they have no plans to replace this property in the immediate future. On the basis of these facts, you conclude that the excess operating cost penalty will continue for the remainder of this asset's physical life, nine years.

Assuming a 10 percent discount rate and a 40 percent income tax rate, the quantification of operating obsolescence is as shown in table 3.17.

Annual Excess Operating Costs	$12,500
Less Taxes at 40%	$5,000
Annual Excess Operating Costs After Tax	$7,500
Present Value Factor of 10% for 9 Years (See Table PVA)	5.75902
Operating Obsolescence: Energy	$43,193
Rounded	$43,000

Table 3.17. Estimation of Operating Obsolescence

Thus, using the cost approach, the fair market value in continued use of Model A, the subject property in this example and example 2, is $52,800, which is developed as shown in table 3.18. (Note that this assumes there is no economic obsolescence.)

Reproduction Cost New	$130,000	
Replacement Cost New		$160,000
Less Physical Deterioration: 40%		–$64,000
Replacement Cost New Less Physical Deterioration		$96,000
Less Functional Obsolescence From Excess Operating Costs		–$43,000
Replacement Cost New Less Physical Deterioration and Functional Obsolescence		$53,000
Less Economic Obsolescence		–$0
Fair Market Value in Continued Use		$53,000

Table 3.18. Fair Market Value in Continued Use

Example 11 raises several points requiring further discussion. For purposes of illustration, we have assumed that the annual excess operating cost of $12,500 will remain constant during the nine remaining years; that is, we are assuming that production and the cost of energy will remain constant. In the real world, it is often the case that operating cost differentials do not remain constant. If they do not remain constant, the appraiser will need to determine the present value of each year's operating cost differential. Furthermore, in theory the discount rate should reflect the assumptions, including the risk, inherent in the projections and income stream that are the basis for the operating cost differential, which in example 11 would be the projections of future energy costs and production. If the risk inherent in these projections is higher or lower than the risk inherent in the subject plant's normal operations, the discount rate used to estimate operating obsolescence should be increased or decreased to reflect the different level of risk.[27]

There can also be a problem when analyzing operating costs of facilities operating at significantly reduced capacities. Those elements that contribute to operating obsolescence are known as *variable operating costs*; that is, they vary with production. For example, to produce a given product, "x" amount of energy, "y" amount of labor, and "z" amount of raw material are required. At or near full capacity there is a linear (direct) relationship between the amount of input and the amount of output. However, at some lower operating level, the relationship becomes distorted and is no longer linear for all elements. This results in greater operating costs per unit of output. Further discussion of this topic is beyond the scope of this book, but the appraiser should be aware that distortions can occur.

Example 11 illustrates a situation in which the higher capital cost of the modern replacement asset is more than offset by its lower operating cost, as compared to the less efficient subject asset. In example 11 there were sufficient data to draw an objective conclusion of value, using the cost approach, for that particular asset. However, in some cases there may be a dramatically higher capital cost without any offsetting savings in operating cost. This is especially true when appraising properties such as pollution abatement facilities that have been affected by a change in pollution control regulations. For example, consider an appraisal of a dust collection system that was designed for particulate emission of 100 PPM (parts per million), where the facility operates within its designed capability. Assume that a new government regulation states that all new dust collection systems can emit only 50 PPM, that this requirement effectively doubles the capital cost, and that it will also require higher energy and labor cost to meet the new requirement. Furthermore, assume that the existing dust collection systems, including the one being appraised, will remain in operation under a "grandfather" clause in the regulation. The question here is: What is the proper level of current cost for the dust collection system being appraised? The answer is the reproduction cost new of the subject facility, because the specifications for the modern facility are far greater than those for the subject facility. In fact, in this example, there is no modern equivalent because of the regulations. In addition, a prudent investor is not going to build the newer dust collection system without realizing some economic benefit (for example, reduced costs and/or improved quality), or without being legally required to do so, as in this case.

Economic Obsolescence

The last step in the implementation of the cost approach is to estimate *economic obsolescence*. Economic obsolescence (sometimes called "external obsolescence") has been previously defined as the loss in value or usefulness of a property caused by factors external to the asset. These factors include increased cost of raw materials, labor, or utilities (without an offsetting increase in product price); reduced demand for the product; increased competition; environmental or other regulations; or similar factors.

The difficulty in measuring the full effect of economic obsolescence is one of the weaknesses of the cost approach. Because economic obsolescence is usually a function of outside influences that affect an entire business (i.e., all tangible and intangible assets) rather than individual assets or isolated groups of assets, it is sometimes measured using the income approach or by using the income approach to help identify the existence of economic influences on value. However, the cost approach can be used to measure some forms of economic obsolescence.

Inutility

The cost-to-capacity concept, discussed earlier in this chapter, can be used to estimate one form of economic obsolescence within the cost approach. Whenever the operating level of a plant or an asset is significantly less than its rated or design capability, and the condition is expected to exist for some time, the asset is less valuable than it would otherwise be. An *inutility penalty* can be used to measure the loss in value from this form of economic obsolescence.

There are at least two methods of measuring this loss of value using the cost approach. The first method, which is illustrated below, assumes there is no fixed expense associated with the plant or production line of which the subject is a part. This is an unrealistic assumption in most situations. Unless fixed expense is zero, the method illustrated below may significantly underestimate the inutility penalty. If the plant's or production line's fixed expense is not zero, the second method should be considered.[28]

The first method measures the loss in value by reducing the capital investment from rated capability to the actual operating level to "balance" the plant. For example, assume that the task is to appraise a property that has a rated or design capacity of 1000 tons/day, but it is operating at only 750 tons/day because of reduced demand for the product or other unfavorable external conditions. If the replacement cost new is developed based on 1000 tons/day and the operating cost penalties are discounted at 750 tons/day, an obvious imbalance exists. That imbalance is the unused capacity reflected in the replacement cost new estimate but not reflected in the operating obsolescence. This unused or unproductive capacity should be reflected in your depreciation estimate.

The inutility penalty can be calculated on a percentage basis by comparing the actual operating level to the rated capacity using the following formula:

$$\text{Inutility as a Percent} = \left[1 - \left(\frac{\text{Capacity } B}{\text{Capacity } A}\right)^n\right] \times 100 =$$

$$(1 - 0.818) \times 100 =$$

$$0.182 \times 100 = 18.2 \text{ percent}$$

Where capacity A	=	rated or design capacity
capacity B	=	actual production
n	=	exponent or scale factor

(Note: If the exponent is not available from published sources, it can be determined as shown in example 6.)

This formula is based on the cost-to-capacity concept discussed earlier in this chapter, whereby the cost of facilities of different capacities vary exponentially rather than linearly because of economies of scale. In other words, as capacity increases, cost also increases but at a different rate. This same logic is used to develop the inutility penalty. As discussed earlier, scale factors vary depending on the type of equipment and labor/material ratios.

It is important to note that the use of this method applies to both functional and economic penalties, depending on the cause of inutility. Again, the primary purpose is to balance the plant from both a capital and operating cost basis. Continuing with the example given above, let us say that you discount the operating obsolescence to present value based on 750 tons/day. The actual inutility is the 250 tons/day that is not operating. Under the principle of substitution, a prudent investor would not purchase this unproductive capacity without being able to realize some benefit. If the plant is not operating at capacity for economic reasons, the inutility is caused by economic obsolescence. If there is an imbalance in the productive capacity (e.g., a production bottleneck), the inutility is caused by functional obsolescence.

Finally, although it is not common, if a property is not operating at capacity because of some physical problem, the inutility is caused by physical deterioration. Once again, the appraiser should determine the facts and circumstances and apply them as appropriate.

Example 12: Inutility—measuring economic obsolescence within the cost approach.

You are appraising a production line capable of 1,000 units per day. It is approximately three years old and is in excellent condition. The production line represents the state of the art from a technological point of view. In your discussions with the plant manager, you learn that there has been a dramatic increase in foreign competition and, consequently, the client is operating the line at only 750 units per day.

For this example, assume that the replacement cost new for a plant with a capacity of 1,000 units per day is $1,000,000; that the proper scale factor for this kind of facility is 0.7; and that you already determined that physical deterioration is 15 percent. Your task is to measure the additional depreciation (obsolescence) attributable to the reduced production.

The reduced operating level is an element of economic obsolescence since it is caused by factors external to the property. In this case, the application of an economic inutility penalty is appropriate and is developed as follows:

$$\text{Inutility as a Percentage} = \left[1 - \left(\frac{750}{1{,}000}\right)^{0.7}\right] \times 100 =$$

$$(1 - .818) \times 100 =$$

$$0.1820 \times 100 = 18.2 \text{ percent}$$

The determination of the fair market value in continued use of this asset, using the cost approach, is summarized as shown in table 3.19.

Replacement Cost New	$1,000,000
Less Physical Deterioration at 15% (given)	–$150,000
Replacement Cost New Less Physical Deterioration	$850,000
Less Functional Obsolescence From Excess Operating Costs	–$0
Replacement Cost New Less Physical Deterioration and Functional Obsolescence	$850,000
Less Economic Obsolescence Calculated at 18.2%	–$154,700
Fair Market Value-In-Place in Use	$695,300
Rounded	$700,000

Table 3.19. Measuring Inutility within the Cost Approach

In the preceding example, there are several points worth discussing. First, notice that the inutility penalty is not linear: a 25 percent decrease in operating capacity results in an 18.2 percent inutility. Second, the penalty was deducted after physical deterioration and functional obsolescence (there was no functional obsolescence in this example) because economic obsolescence is independent of physical deterioration and functional obsolescence. Third, a valid question can be raised regarding the proper capacity to use as the basis for developing replacement cost new. If the economic conditions are such that the long-term production will be 750 units per day, it may be valid to calculate replacement cost new based on a 750-unit-per-day plant instead of the 1,000-unit-per-day plant used in the example.

Developing an inutility penalty is a way of measuring one form of economic obsolescence within the cost approach. In practice, when dealing with relatively new assets that are not operating at their capacity because of economic reasons, additional economic obsolescence is probably present. To measure this may require a detailed analysis of the business and a subsequent allocation of any economic penalties to the individual assets or groups of assets.

Other Methods of Measuring Economic Obsolescence Within the Cost Approach

Unfavorable external factors may cause the subject plant or asset to incur excess operating costs. These excess operating costs, caused by external instead of internal factors, can be measured and converted into an economic obsolescence penalty using the same methods discussed under operating obsolescence (see examples 10 and 11).

An example would be the significantly higher raw material costs experienced by paper mills and other forest products facilities in the Pacific Northwest in the 1990s. Higher raw material costs were caused primarily by court decisions in environmental cases that had the effect of substantially restricting the supply of raw material, thereby raising raw material costs for paper mills and other forest products mills in the Pacific Northwest. The raw material costs of paper mills in other parts of the United States, with which the Pacific Northwest mills competed, were not affected by the court decisions. Since the Pacific Northwest mills were not able to pass on the increased cost to their customers (because the paper market is national or global), these mills experienced substantial excess operating costs, when compared to mills with which they competed that were not affected by the court decisions. These excess operating costs, caused by external forces, could be measured and converted into an economic obsolescence penalty using the same methodology presented in our discussion of operating obsolescence.

Other methods are available for quantifying economic obsolescence within the cost approach. While a discussion of these methods is beyond the scope of this book, it should be noted that other measures of economic obsolescence can be developed based on analyses of industry returns, supply/demand relationships, margin analysis, product or raw material price changes, stock prices, the relationship between replacement cost new and the cash flows the hypothetical replacement facility is capable of generating,[29] and other economic evidence indicating that the value of the subject property has been reduced by external factors. As long as appraisers objectively examine the facts, apply appropriate analytical methods, and avoid double-counting depreciation, they are limited only by their resourcefulness and creativity.

Sequence of Depreciation

We have discussed the cost approach using a very distinct sequence or order: current cost new less physical deterioration, functional obsolescence, and economic obsolescence. This is the generally accepted method for deducting the elements of depreciation from the cost new.

The traditional sequence of the cost approach is as follows:

Step 1:	Reproduction Cost New Less Excess Capital Cost = Replacement Cost New
Step 2:	Replacement Cost New (RCN) Less Physical Deterioration = RCN Less Physical Deterioration (RCNLPD)
Step 3:	RCNLPD Less Functional Obsolescence = RCNLPD and Functional Obsolescence (RCNLPD + FO)
Step 4:	RCNLPD + FO Less Economic Obsolescence = Replacement Cost New Less All Forms of Appraisal Depreciation

The following is an example of how the traditional cost approach sequence of depreciation is applied.

Assume that the following has already been established by the appraiser:

Reproduction Cost New	=	$150,000
Excess Capital Cost	=	– $10,000
Replacement Cost New	=	$140,000
Physical Deterioration	=	50 percent
Functional Obsolescence	=	20 percent
Economic Obsolescence	=	$16,000

The usual cost approach calculations would then be as follows:

Reproduction Cost New	=	$150,000
Less Excess Capital Cost	=	– $10,000
Replacement Cost New	=	$140,000
Less Physical Deterioration @ 50%	=	– $70,000
		$70,000
Less Functional Obsolescence @ 20%	=	– $14,000
		$56,000
Less Economic Obsolescence	=	– $16,000
Replacement Cost New less all Depreciation		$40,000

The above reflects the traditional methodology for the cost approach. It is important to recognize that after each calculation a new subtotal is derived. All of the forms of depreciation—physical deterioration, functional obsolescence, and economic obsolescence—are deducted in the order and manner shown above. They are *not* added together or combined and then deducted from cost new.

The logic behind the sequence is derived from the normal life cycle of a property. When a property is new, its value is usually equal to the sales price. There are a willing buyer and seller, implying that an economic justification (a business need of some kind) for the purchase of the property exists. Once the property is placed in service by the buyer, it begins to depreciate. Usually the first element of depreciation that occurs is physical deterioration, since the property will probably be used for the purpose for which it was purchased. As the property continues in operation, two elements of deterioration come into play, curable and incurable. Curable deterioration is usually correctable through routine maintenance, while incurable deterioration begins in the form of metal fatigue (or some similar condition) and cannot be remedied. Physical deterioration is the only element of depreciation that affects the item until something happens in the marketplace or environment to trigger functional or economic obsolescence. Usually an equipment manufacturer will improve the asset gradually over time, so when the manufacturer announces a "new and improved" version of the asset, obsolescence is usually introduced into the existing asset. Normally the new version incorporates some technological improvements, suggesting that functional obsolescence affects the value of the existing asset. At this point the asset is operating, is physically deteriorating, and now exhibits some functional obsolescence. As time goes on, external factors such as reduced profitability in an industry, increased competition, foreign imports, a shift in market demand, or new government regulations cause the item to experience some form of economic obsolescence. Thus, economic obsolescence is usually the last element of depreciation to affect the property.

This is the normal sequence for calculating and applying depreciation when using the cost approach. The rationale for calculating depreciation in this sequence is that it, in most instances, accounts for depreciation in the order that it occurs. However, there may be occasions when the facts or economic logic will dictate that the sequence of deductions from cost new should be different. It should also be noted that when a deduction is made for excess capital costs (the difference between reproduction and replacement cost new), one is deducting a form of obsolescence before considering any other forms of depreciation. The most important point, however, is that when using the cost approach, an appraiser should strive to segregate the various elements of depreciation and avoid doubling-up on or omitting some form of depreciation.

Key Points

- Using the cost approach, the appraiser starts with the current replacement cost of the property being appraised and then deducts for the loss in value caused by physical deterioration, functional obsolescence, and economic obsolescence. The logic behind the cost approach is the principle of substitution: a prudent buyer will not pay more for a property than the cost of acquiring a substitute property of equivalent utility. The principle can be applied either to an individual asset or to an entire facility.
- The replacement cost new is the proper starting point for developing an opinion of value using the cost approach.
- It is essential that the appraiser understand the difference between *replacement cost* and *reproduction cost*. Replacement cost is the current cost of a property with utility equivalent to the subject, whereas reproduction cost is the current cost of a replica of the subject. When using the cost approach, the appraiser is comparing the property that would replace the subject to the subject property. The replacement property would be the most economical new property that could replace the service provided by the subject.
- There are several methods of determining the current new cost of a property. The major ones are the *detail method*, *trending*, *cost-to-capacity*, and *other engineering methods*.
- The *detail method,* also known as the summation method, requires that a current new cost be assigned to each individual component of a property. The property is itemized or "detailed" so that the sum of the components reflects the cost new of the whole.
- *Trending* is a method of estimating a property's reproduction cost new (not replacement cost new) in which an *index* or *trend factor* is applied to the asset's *historical cost* to convert the known cost into an indication of current cost.
- *Appraisal depreciation* has been defined as the difference in value between an existing old asset and a new asset taken as a standard of comparison. Appraisal depreciation should measure value inferiority.

- The three types or causes of appraisal depreciation traditionally recognized by appraisers are *physical deterioration, functional obsolescence,* and *economic obsolescence.*
- *Physical deterioration* is the loss in value or usefulness of a property due to the using up or expiration of its useful life caused by wear and tear, deterioration, exposure to various elements, physical stresses, and similar factors.
- *Functional obsolescence* is the loss in value or usefulness of a property caused by inefficiencies or inadequacies of the asset itself, when compared to a more efficient or less costly replacement asset that new technology has developed. Symptoms suggesting the presence of functional obsolescence are excess operating cost, excess construction (excess capital cost), over-capacity, inadequacy, lack of utility, or similar conditions.
- *Economic obsolescence* (sometimes called "external obsolescence") is the loss in value or usefulness of a property caused by factors external to the property, such as increased cost of raw materials, labor, or utilities (without an offsetting increase in product price); reduced demand for the product; increased competition; environmental or other regulations; or similar factors.
- Three methods of measuring physical deterioration that were discussed are *observation, formula/ratio,* and *direct dollar measurement.*
- In the *observation method,* the appraiser makes a comparison based on the experience he or she has gained by looking at similar properties and comparing them to new properties.
- In one variation of the *formula/ratio method,* physical deterioration is estimated based on a property's use. Use is a good indicator of physical deterioration when the requisite production statistics can be obtained.
- The *age/life* variation of the formula/ratio method uses the ratio of a property's "age" to its "life" to measure physical deterioration. Although this is straight-line depreciation, it should not be confused with accounting depreciation, because the appraiser uses valuation concepts of age and life.

- Functional obsolescence resulting from *excess capital costs* is measured by the difference between reproduction and replacement cost (assuming replacement cost is lower). This excess cost represents the decreased capital investment required to obtain the most economical new asset to perform the same service as the subject. Functional obsolescence due to excess capital costs results from improvements that make the new technology more desirable, such as changes in design, materials, layout, product flow, construction methods, and equipment size and mix.
- A second type of functional obsolescence is that caused by *excess operating costs.* As a result of new technology, in many cases not only it is cheaper to acquire a modern replacement asset (capital cost), but it is also cheaper or more efficient to operate the new asset (operating cost). Calculating operating obsolescence involves a comparison of the operating characteristics of the subject property to its modern equivalent, the property with which the subject would (hypothetically) be replaced. The reduced costs being achieved by the modern replacement property are compared to the existing property or facility and its higher operating costs. From this comparison, a basis for estimating a penalty for continued use of the existing property is developed.
- The difficulty in measuring the full effect of economic obsolescence is one of the weaknesses of the cost approach. Because economic obsolescence is usually a function of outside influences that affect the entire business (i.e., all tangible and intangible assets) rather than individual assets or isolated groups of assets, income approach methods are sometimes used to estimate economic obsolescence. However, the diligent and resourceful appraiser can usually find a way to apply the cost approach to most, if not all, forms of obsolescence.
- The *cost-to-capacity* concept can be used to estimate one form of economic obsolescence within the cost approach. Whenever the operating level of a plant or an asset is less than its rated or design capability, the asset is less valuable than it would otherwise be. An *inutility penalty* can be used to measure the loss in value.

- Unfavorable external factors may cause the subject plant or asset to incur excess operating costs. These excess operating costs, caused by external instead of internal factors, can be measured and converted into an economic obsolescence penalty using the same methods presented in our discussion of operating obsolescence.

Additional Reading

Appraisal Institute. *The Appraisal of Real Estate*. 11th ed. Chicago: Appraisal Institute, 1996.

Babcock, Henry A., FASA. *Appraisal Principles and Procedures*. Washington, DC: American Society of Appraisers, 1989.

Bonbright, James C. *The Valuation of Property*. Reprint, Charlottesville, VA.: The Michie Company, 1965.

Grant, Eugene L., and Paul T. Norton, Jr. *Depreciation*. Revised printing, New York: The Ronald Press Company, 1955.

Humphreys, K., and S. Katell. *Basic Cost Engineering*. New York: M. Dekker, 1981.

Marston, Anson; Robley Winfrey; and Jean C. Hemstead. *Engineering Valuation and Depreciation*. Iowa City: Iowa State University Press, 1953.

Notes

[1] *ARE,* p. 336; James C. Bonbright, *The Valuation of Property,* p. 157.

[2] See Bonbright, *The Valuation of Property;* Grant and Norton, *Depreciation;* and Svoboda, Robert S., "Fair Market Value Concepts." In *Appraising Machinery and Equipment.* Herndon, VA.: American Society of Appraisers, 1989.

[3] Ibid.

[4] Bonbright, *The Valuation of Property,* 1965, p. 163; bracketed material has been added.

[5] Grant and Norton, *Depreciation,* p. 277.

[6] Replacement cost new is often less than reproduction cost new, but this is not always the case (see Example 2).

[7] A 10% per year trend factor is used for illustration purposes only; in the real world, annual trend factors are not usually this high.

[8] *ARE,* p. 347.

[9] The definitions of historical cost and original cost used in this book are consistent with accepted appraisal terminology (see the *Dictionary of Real Estate Appraisal,* 2nd ed., pp. 150 and 216); but these terms are often used interchangeably by accountants and others. The appraiser must often discriminate between the two terms, because used equipment is often the subject of an appraisal or a sales comparison analysis. Such cases require the appraiser to know if the cost represents a historical or original cost, or both.

[10] *Webster's New World College Dictionary,* 3rd ed. (New York: MacMillan USA), p. 686.

[11] *ARE,* 11th ed., p. 351.

[12] *ARE,* p. 351.

[13] K. Humphreys and S. Katell, *Basic Cost Engineering,* p. 12.

[14] Humphreys and Katell, *Basic Cost Engineering,* p. 12; Merritt Agabian, ASA, "Replacement Cost New Concepts," *Appraising Machinery and Equipment,* Herndon, VA.: American Society of Appraisers, 1989, p. 45; Nobel L. Davis, "Machinery and Equipment Trends—How Are They Used?" *Valuation*, vol. 17, no. 2 (1970), p. 64.

[15] Frederic C. Jelen and James H. Black, *Cost and Optimization Engineering* (New York: McGraw Hill, Inc., 1983).

[16] For example, see Table 14.15 at p. 348 of Jelen and Black, *Cost and Optimization Engineering*; and Table 2.4 at pp. 18–19 of Humphreys and Katell, *Basic Cost Engineering.*

[17] Jelen and Black, *Cost and Optimization Engineering.*

[18] Chilton, C.H., *Cost Engineering in the Process Industries* (New York: McGraw-Hill, 1960).

[19] Humphreys and Katell, *Basic Cost Engineering,* pp. 12–22.

[20] Grant and Norton, p. 268.

[21] Grant and Norton, pp. 269–270; see also Bonbright, pp. 185 and 189. Note that the

substitute asset may not be hypothetical and often actually exists.

[22] This conclusion assumes that the services of the asset are worth the same in its later period of use as in its early period of use, which may not be the case due to increases in operating costs or reductions in operating efficiency with age; see the discussion of this point at pages 235–236 of *Engineering Valuation and Depreciation,* Marston et al.

[23] Anson Marston, Robley Winfrey, and Jean C. Hemstead, *Engineering Valuation and Depreciation* (Iowa City: Iowa State University Press, 1953)

[24] See example 2.

[25] In this discussion, the term *operating costs* is intended to refer to total manufacturing costs, which would include most elements of "cost of goods sold" as well as "operating expenses."

[26] This may be a different discount rate than the plant's or company's weighted average cost of capital. For example, see *Les Schwab Tire Centers of Oregon v. Crook County Assessor,* Oregon Tax Court, where the court ruled that a riskless rate should be used to discount the future excess operating costs because the riskiness of the excess operating costs was less than the overall cash flows produced by the business.

[27] See footnote 26.

[28] For a detailed description of the second method, see Robert G. Crawford & Gary C. Cornia, "The Problem of Appraising Specialized Assets," *The Appraisal Journal*, Jan. 1994, pp. 75–85.

[29] With respect to economic obsolescence on an overall plant basis, if new plants are not being built, it may be because the cost new of replacement plants is higher than the present value of the cash flows that new plants could produce. Thus, if new plants are not being built, it may be an indication of significant industry-specific economic obsolescence. In such a situation it may be that use of the cost approach, at least on an overall plant basis, is not appropriate.

4

Sales Comparison Approach

Objectives:

1. Describe the sales comparison approach.
2. Discuss elements of comparability.
3. Illustrate the application of the sales comparison approach to single items, production lines, and whole plants.

The MTS appraiser uses the sales comparison approach[1] to indicate value by analyzing recent sales (or offering prices) of properties that are similar (i.e., comparable) to the subject property. If the comparables are not exactly like the properties being appraised, the selling prices of the comparables are adjusted to equate them to the characteristics of the properties being appraised. The basic procedure is to gather data on sales and offerings of similar properties, determine their comparability to the subject property, determine the appropriate units of comparison, collect and array the data, analyze and adjust the data, and apply the results to the subject. Like the cost and income approaches, the sales comparison assumes that the informed purchaser would pay no more for a property than the cost of acquiring a comparable property with the same utility.

This approach focuses on the actions of actual buyers and sellers. In theory, the approach measures the loss in value from all forms of appraisal depreciation that are inherent in the individual asset, assuming appropriate adjustments are made to the comparables to reflect differences between them and the subject.[2]

The used equipment market is an established means of buying and selling equipment. The used market consists of used machinery dealers, auctions, and public and private sales, and is often (but not always) the most reliable method of determining certain types of value for certain types of properties.

The sales comparison approach is most reliable when there is an active market providing a sufficient number of sales of comparable property that can be independently verified through reliable sources. Examples of properties generally having such markets are automobiles and trucks, computers, aircraft, and other properties with an identifiable market. The important concepts are "active market" and "verifiable information." An active market has truly independent transactions occurring under free market conditions. When researching market sales, the appraiser should verify that the sales are independent rather than being conducted by one seller or buyer (the latter situation could create a false appearance of an active market). There is no set number of sales that make a market.

The sales comparison approach is not feasible when the subject property is unique. Even if the subject property is not unique, the approach will generally not be feasible if an active market for that property does not exist. An inactive market, or one where there are a limited number of sales of comparable property, often indicates a lack of demand and the existence of economic obsolescence: where an inactive market exists, property might be better analyzed using the income or cost approaches.[3]

This chapter discusses the sales comparison approach as applied to an individual asset (such as a single item of equipment), a related group of assets (such as a production line), and an entire industrial facility. The implementation of the sales comparison approach may differ significantly depending on whether the subject is an individual asset, a group of assets, or an entire facility. The approach generally becomes more complicated when applied to a group of assets or an entire facility, because buyers (and by implication sellers) of these more substantial assets often, either explicitly or implicitly, consider the present value of the future benefits of ownership (e.g., net cash flow) in making purchasing decisions. Although this consideration reflects an income approach, this concept must also be taken into account in the sales comparison approach, since all valuation methods must attempt to replicate the analysis and behavior of buyers and sellers in the real world.

Premises of Value

The appraiser's analysis begins, not with the search for comparables, but with the determination of both the appraisal's purpose and its appropriate premise of value. It is essential to determine the proper value premise at the beginning of the valuation assignment. Different premises may require consideration of different facts. The following premises of value applicable to the sales comparison approach are the same as those discussed in chapter 1:

- fair market value;
- fair market value in continued use;
- fair market value–installed;
- fair market value–removed;
- orderly liquidation value;
- forced liquidation value; and
- liquidation value in place.

These premises of value embody fundamental concepts and various level-of-trade considerations. Certain appraisals may require variations of these definitions. For example, leases, contracts, regulations, laws, or court decisions may require certain variations; or different appraisers or appraisal firms may have a preference for certain variations. Thus, it may be appropriate to modify the definitions of the above terms[4] to match the purpose or use of a particular appraisal, but the appraiser should be careful not to alter the fundamental concepts embodied in these definitions without compelling reasons.

Identification of the Subject Property

One of the first steps in the sales comparison approach is the proper identification of the subject asset. The microidentification of the subject property is discussed in detail in chapter 2. Microidentification includes determining and listing such characteristics as

- make,
- model,
- serial number,
- size,
- capacity,
- year of manufacture,
- attachments, and
- condition.

Comparable Sales and Adjustments

Recent sales of assets *identical* to the subject often cannot be found. If so, it is necessary to find sales of assets providing comparable or equivalent utility. It should be understood that "comparables" will often be just that: comparable but not identical to the subject.

If the comparable sale is not identical to the subject, the selling price of the comparable must be adjusted to indicate what the selling price of the comparable would have been if the comparable had been identical to the subject. The appraiser should remember that adjustments are made to the comparables, not to the subject property. Adjustments are made for differences between the comparable's and subject's condition, capacity, size, effective age, date of sale, circumstances of sale (level of trade or to a dealer, "as-is, where-is" condition, etc.), location, environmental compliance, safety compliance, and other factors that would have affected the sale price of the comparable.

When adjusting a comparable sale, the appraiser is determining how much more or how much less the comparable would have sold for if it had been identical to the subject in a given single characteristic, such as effective age. For example, if the comparable's effective age was ten years, compared to the subject's effective age of five years, the appraiser would normally make an upward adjustment to the comparable's actual selling price (i.e., increase the comparable's selling price) to reflect the appraiser's opinion of what the comparable's selling price would have been if its effective age (when it sold) was five years instead of its actual effective age of ten years.

When appraising under the concept of *continued use* or *installed,* the appraiser will generally adjust the comparables to include the direct and indirect installation costs. Similarly, the appraiser will adjust the comparables to reflect a different type of sale: for example, if the comparable is a sale to a dealer, an adjustment will be made to equate that sale with the appropriate level of trade for an *installed* or *continued use* premise that is being applied to the subject property.

Unlike real estate appraisal, the machinery appraisal generally does not inspect each comparable. This would be impractical in most cases because of the large number of individual assets being appraised and the fact that comparables may come from a large geographical area. The appraiser often uses databases organized by equipment category and covering sales in a large geographical area. It is important that this market data come from reliable sources, which can be verified for accuracy where possible. This does not mean that every comparable sale must be verified, but it does mean that a professional appraiser should use databases that are generally reliable and that provide an understanding of the transaction basis of the comparable sale (e.g., sale to a dealer, from a dealer).

Comparable sales are not the only value indicators an appraiser may use. Current offerings or listings may also be considered.

In and of itself, the number of comparable assets that are currently available in the used market may have a bearing on the value of the subject. If many comparables are being offered for sale, prices may be depressed and there may be little demand for the subject property.

The appraiser should become familiar with the market applicable to the subject property. This market may be local, regional, national, or in some instances international.[5] The international market may require consideration when older production equipment is sold to users in developing countries. Equipment that is obsolete or unable to be operated competitively in the United States may be profitably used in developing economies where there are lower labor, raw material, or other costs.

Elements of Comparability

Ideally, when appraising machinery and equipment using the sales comparison approach, the appraiser should strive to base conclusions on sales of identical assets that have been exchanged in the marketplace. Unfortunately, it is rare to find sales of units identical to the subject. In practice, the market investigation will probably reveal sales of assets that are similar but not identical, and it is this analysis of similarity upon which the appraiser should base an opinion of value. Some of the elements of comparability are the following:

Vintage and Effective Age: The appraiser should try to determine the effective age of the comparable. This usually requires comparing both the vintage and reported condition of the comparable and making adjustments to account for upgrades and rebuilds.

Condition: Differences in condition affect selling prices of similar assets. This is a difficult area of comparison because, while the condition of the subject is known, it is often difficult to ascertain the condition of the comparable. If possible, there should be an investigation into the condition of the comparable.

Capacity: Ideally, the comparable should have the same, or very similar, capacity as the subject. If not, it may be necessary to adjust the comparable's selling price to account for capacity differences.

Features (accessories): The appraiser should strive to compare the subject to comparables with the same features and accessories.

Location: The geographical location of the comparable sale can affect the selling price. In addition, the physical location of an asset within a plant may also affect the selling price. For example, two identical package boilers, one on the main floor and one on the third floor, would be expected to have different selling prices (all other things being equal) because the one on the third floor will require higher dismantling and moving costs.

Manufacturer: The appraiser should try, if possible, to compare the subject to sales of similar assets made by the same manufacturer. If data from the same manufacturer are not available, the appraiser should compare the subject to units manufactured by a company that market participants consider comparable to the manufacturer of the subject property.

Motivation of Parties: This is an important item of comparison, especially for larger units. Appraisers should ask themselves what is the motivation of the buyer and seller and how does their motivation affect the value of the subject. In most cases the selling price of a comparable will differ, for various levels of trade, depending on whether it is purchased by a dealer (for profitable resale) or by an end user (for immediate installation in a facility). If same-manufacturer data are not available, the appraiser should compare the subject to units manufactured by a company that market participants consider comparable to the manufacturer of the subject property.

Price: In many cases, especially with larger properties, the transaction price should be investigated and expressed on a cash basis. This is particularly true if favorable financing or a trade-in was involved in the comparable-sale transaction. This is commonly called a cash equivalency adjustment.

Quality: The quality of the comparable should be equivalent to the subject. If not, the appraiser should either discard the comparable or make an appropriate adjustment.

Quantity: Unit prices can vary considerably depending on the quantity sold. Adjustments must be made for bulk or large-quantity sales. Quantity is also related to market conditions: a buyer's market suggests that a larger quantity is available, while a seller's market suggests a limited quantity.

Time of Sale: The appraiser should strive to obtain sales occurring within a reasonable period of time from the appraisal's effective date. This is especially important during volatile markets. In theory, comparable sales should be close to the effective date of the appraisal, but these are not always possible to obtain. When sales that occur beyond "a reasonable" period of time need to be considered, the appraiser should explain this and make appropriate adjustments if the data are less than desirable.

Type of Sale: The type and terms of sale generally indicate different price levels or levels of trade. The same asset that is purchased by a machinery dealer at an auction (usually a *liquidation* premise) will probably have a higher price when it is sold by the dealer to an end user (a *market value installed* or *market value in continued use* premise). The dealer is purchasing for profitable resale while the end user is purchasing for immediate installation at the facility. The end user's other option is to purchase from the dealer or another end user, both of which require that the seller be compensated above the level of a liquidation sale as motivation to sell.

Techniques of Comparison

The following are the three most commonly used techniques for establishing value of individual items of machinery and equipment using the sales comparison approach:

Direct Match: This technique establishes value based on a direct match of the subject to an identical asset or comparable. A good example is an automobile valued using a published pricing guide. If the manufacturer, model number, age, and accessories are known, it is relatively simple to determine the value of the subject. Adjustments are limited to mileage and, more important, condition. In this case, the appraiser is directly comparing the subject to the compilation of sales of other identical autos. The direct match technique is certainly nothing new, but it is identified as a separate technique because it provides what is probably the most accurate indication of value using the sales comparison approach. Without a direct match, value conclusions become more subjective.

Comparable Match: This technique establishes value based on analysis of similar (but not identical) assets using some measure of utility (size, capacity, etc.) as the basis of comparison. For example, when appraising an engine lathe manufactured by Company A, the appraiser finds no sales of similar engine lathes manufactured by Company A but does find sales of similar engine lathes manufactured by Companies B and C. Obviously, this technique is more subjective than a direct match, requiring additional adjustments based on an analysis of the elements of comparability discussed earlier. For example, the appraiser would have to judge whether the typical market participant would consider engine lathes manufactured by Companies A, B, and C to be approximately equal in value.

Percent of Cost: This technique is nothing more than establishing the ratio of the selling price to the current cost new of a property at the time of sale. With sufficient data, similar properties can be statistically analyzed and relationships developed among age, selling price, and cost. For example, an appraiser is valuing a 16"× 208" engine lathe manufactured by Company A. The market investigation does not find a direct match. It does identify many somewhat similar lathes manufactured by different companies (including Company A), but the sizes of these lathes are either much smaller or much larger than the subject lathe. Assuming the analysis suggests that selling prices of engine lathes with an age and condition similar to the subject are in the range of 40 to 50 percent of current cost new, it would probably be logical to conclude that the subject's value falls somewhere between 40 and 50 percent of its cost new. It should be noted that the market for a unit may vary according to the unit's size. For example, small lathes may appeal only to maintenance shops, medium-sized lathes may appeal to standard machine shops, while very large lathes may be used only in oilfield, shipyard repair, or railroad applications. Appraisers should ensure that their data set fits their subject.

Appraising an Individual Unit

The following example applies the sales comparison approach to an individual unit, such as a single piece of equipment. Some sources of market data used for appraising a single piece of equipment are

- used equipment dealers or other sellers,
- used equipment buyers,
- equipment databases,
- auction sales databases,
- Internet databases,
- client fixed asset ledgers,
- trade publication classifieds,
- leasing companies, and
- other appraisers.

Again, adequate identification of the subject is necessary before implementing the sales comparison approach. A sample description for a hypothetical single piece of equipment is shown below.

Type of Equipment	Crawler loader
Manufacturer	XYZ Industries
Model	CT4
Serial Number	CT478
Year Manufactured	1990
Observed Condition	Very good
Description	Low ground pressure model with a 6-way blade, rollover protection system, diesel engine, very good undercarriage
Location	Houston, Texas
Effective Date of Appraisal	Current Date

This description is an example of one way to list a piece of equipment. There are other acceptable formats. Some may include certain attachments or appurtenances. Once the subject is described, the process of identifying market sales begins.

Liquidation Value—Individual Unit

If the premise of value were *liquidation value* for the crawler loader described above, the first step would be to search a number of data sources for comparable sales at a liquidation level of trade. Data sources might include publications by *Equipment World* and *Top Bid* that cover transactions relating to construction equipment, trucks, and trailers. Other possible sources include used equipment dealers and other publications, some of which are listed in the appendix. It may also be helpful to contact other appraisers and dealers who maintain individual databases.

After accurately describing the property and conducting a search for market sales, ten auction sales are found as potential comparables. These are shown in the following list.

1. Description: XYZ Industries M/N CT4 S/N 430 (Good Condition) Low Ground Pressure Model, 6-way Blade, Rollover Protection Structure, Good Undercarriage.

   Transaction: Auctioneers, Inc.
   April 1994, Gadsden, AL
   $54,000 Sale Price

2. Description: XYZ Industries M/N CT4 S/N 414 (Very Good Condition) Low Ground Pressure Model, 6-way Blade, Drawbar, Engine Enclosed, Rollover Protection Structure, 30" Pyramid Pads, Very Good Undercarriage.

   Transaction: Local Auctioneers
   April 1994, Fort Worth, TX
   $50,000 Sale Price

3. Description: XYZ Industries M/N CT4 S/N 444 (Good Condition)
   Low Ground Pressure Model, 6-way Blade, Engine Enclosed, Open Rollover Protection Structure, Single Bar Grousers, Pads, Very Good Undercarriage.

   Transaction: South-Atlantic Auctions
   January 1994, Houston, TX
   $45,000 Sale Price

4. Description: XYZ Industries M/N CT4 S/N 430 (Good Condition)
   Low Ground Pressure Model, 6-way Blade, Rollover Protection Structure, Diesel Engine, 30" Shoes, Good Undercarriage.

	Transaction:	Sun Auction Co. February 1994, Kissimmee, FL $55,000 Sale Price
5.	Description:	XYZ Industries M/N CT3 S/N 325 (Condition Unknown) Engine Enclosed, Canopy w/Sweeps & Rear Screen.
	Transaction:	Complete Auctioneers March 1994, Nashville, TN $48,000 Sale Price
6.	Description:	XYZ Industries M/N CT3 S/N 190 (Good Condition) Low Ground Pressure Model, 6-way Blade, Canopy, 36" Single Bar Grousers, Pads, Fair Undercarriage.
	Transaction:	Family Auctioneers, Inc. June 1994, Surrey, BC, Canada $42,600 Sale Price ($US)
7.	Description:	XYZ Industries M/N CT3 S/N 167 (Good Condition) Low Ground Pressure Model, 6-way Blade, Hydraulic Controls, Engine Enclosed, Rollover Protection Structure, 30" Pads,Good Undercarriage.
	Transaction:	Auctioneers, Inc. February 1994, Fort Worth, TX $51,000 Sale Price

8. Description: XYZ Industries M/N CT3 S/N 146
(Good Condition)
Low Ground Pressure Model, 6-way Blade, Engine Enclosed, Rollover Protection Structure, Canopy with Sweeps, Good Undercarriage.

Transaction: Auctioneers, Inc.
April 1994, Gadsden, AL
$52,000 Sale Price

9. Description: XYZ Industries M/N CT3 S/N 350
(Condition Unknown)
6-way Blade, Rollover Protection Structure, Diesel Engine.

Transaction: Sunn Auction Co.
April 1994, Columbus, OH
$42,000 Sale Price

10. Description: XYZ Industries M/N CT3 S/N 327
(Condition Unknown)
Hydraulic, 6-way Blade, Multishank Rear Ripper, Canopy.

Transaction: West Coast, Inc.
December 1993, Moreno Valley, CA
$42,500 Sale Price

The ten sales are summarized in table 4.1.

	Price	Date of Sale	Condition	Sale
Location				
Sale 1	$54,000	4/94	VG	AL
Sale 2	$50,000	4/94	VG	TX
Sale 3	$45,000	1/94	G	TX
Sale 4	$55,000	2/94	G	FL
Sale 5	$48,000	3/94	UNK	TN
Sale 6	$42,600	6/94	G	CAN
Sale 7	$51,000	2/94	G	TX
Sale 8	$52,000	4/94	G	AL
Sale 9	$42,000	4/94	UNK	OH
Sale 10	$42,500	12/93	UNK	CA

Table 4.1. Comparable Sales Summary

After analyzing the ten comparable sales and checking serial number guides, it is determined that the properties from sales 1 through 4 were manufactured in 1990, the same year as the subject, while the properties from the remaining six sales (5 through 10) were a year older, or 1989 models. Prices range from $42,000 to $52,000 for the 1989 models and $45,000 to $55,000 for the 1990 models.

The next step in the valuation process is to compare these sales to the subject property. For the crawler loaders manufactured in 1990, the first adjustment is made for condition. The subject's condition is "very good." A review of the sales shows that sales 1 and 2 were in "very good" condition, while sales 3 and 4 were in only "good" condition. Thus, the appraiser needs to consider how much higher the prices in sales 3 and 4 might have been if those items' condition had been identical to the subject's. Under the circumstances, an upward adjustment to sales 3 and 4 is probably warranted. But by how much? Quantifying adjustments is one of the most difficult aspects of sales comparison approach appraisals. The best technique, if the data are available, is the "paired sales" technique. The appraiser notes that sales 2 and 3 are identical in every relevant respect except for condition and selling price. Sale 3's condition was "good" and it sold for $45,000; sale 2's condition was "very good" and it sold for $50,000. Sales 2 and 3 are "paired sales" and it would be logical, in the example given, to conclude that sale 3 would have sold for $5,000 more if its condition had been "very good" like the subject's. Even though sale 4 is not identical to sales 2 or 3, it would be appropriate, in the example given, to also adjust its selling price upward by $5,000 to equate its condition with that of the subject.

It is also necessary to investigate the prevailing market conditions when these crawler loaders were sold. Note that the subject is located in Texas. Assuming (for purposes of illustration) (a) that a lack of construction activity in Texas in 1994 limited the demand for crawler loaders, (b) that this situation continues to exist in Texas as of the effective date of the appraisal, and that (c) construction activity was surging in Florida in 1994 when sale 4 was consummated, these circumstances suggest that the price in sale 4 should be adjusted downward to equate the market conditions at the time of this sale with those prevailing at the time of the subject's. Determining the extent of the adjustment will be more subjective than determining the adjustment for condition (illustrated above), because in this case the data do not provide the appraiser with a paired sale (and in the real world this will often be the case). The point here is that the appraiser has investigated the market conditions pertaining to the various sales and realizes that a downward adjustment to sale 4 is needed. The amount of adjustment often will not be precisely quantifiable, but at least the appraiser has done his or her job by investigating those conditions about the market sales that are relevant for valuation purposes.

The appraiser should, whenever possible, contact the auctioneer to determine the conditions surrounding an auction sale. The appraiser should ask questions on the number of people in attendance, the number of active bidders, the weather conditions, the type of advertising, and the number of brochures mailed out. For example, the appraiser would want to know if there were only five people at the auction and only one active bidder, conditions that could have depressed the selling prices. Although it may be difficult to gather this type of information, its absence can impair the reliability of the appraisal conclusions.

This discussion is not intended to illustrate all the adjustments that would be made to arrive at a value conclusion for the subject crawler loader. Rather, it is intended to emphasize that the appraiser's job is to investigate and develop those facts about the market sales and the subject property that are relevant for valuation purposes. Adjustments differ from property to property and from project to project. There are no rules of thumb or specific guidelines that apply in every case. In fact, it is inadvisable to apply the same adjustments to every appraisal assignment. The appraiser cannot simply rely on databases and use information without considering whether adjustments need to be made. The appraiser is primarily an investigator and developer of facts—not all possible facts, but those facts that affect value.

Fair Market Value in Continued Use or Installed—Individual Unit

Consider a situation in which the buyer of a property in liquidation intends to use that asset in his or her industrial operation (a *continued use* or *installed* premise). Assume the buyer purchased for cash on an "as-is, where-is" basis. In this case, the buyer must pay the cost of dismantling and removing the asset, as well as any maintenance or rebuilding costs. If the asset were purchased from a dealer, these costs would normally be "buried," or incorporated, in the dealer's selling price. In theory, the sum of the purchase price plus dismantling, removal, rebuilding, and maintenance costs should be the same for a dealer as for an end user, except for the dealer's overhead and profit.

Example 1: Using sales comparison approach to estimate fair market value in continued use or fair market value installed.

Your task is to estimate the *fair market value in continued use* or *fair market value installed* of an eight-year old milling machine, using the sales comparison approach. The machine is currently being used for custom work, is used frequently, and is maintained on a preventive basis. Observation and discussion with plant personnel confirm that the machine is in good condition.

You attend auctions regularly and have noticed that this particular machine is very popular. The data suggest the machine commands at auction a range of prices from $1,000 to $7,500, depending on age and condition. Recent sales data suggest that, given the age and condition of the subject, it would sell for between $5,500 and $6,500. Discussions with used machinery dealers indicate they would ask $6,500 for this asset and would expect to sell it for $6,000. Based on your knowledge of the market and the confirming discussions with dealers, it is logical to conclude that $6,000 would represent the value of the base unit (i.e., the machine itself).

The next step is to add the installation and other assemblage costs that convert this base unit value amount to the *fair market value installed* (or *fair market value in continued use*). This asset is relatively simple to install and connect. Because it is a common item, assume that the asset would be purchased locally with a freight cost of $200. The time for two millwrights to unload and set the machine in place is two hours for a cost of $125. Electrical installation, including controls, is $300. Therefore, the total cost new of the installation and other assemblage costs is the sum of all of these, or $625.

Since the asset was installed new eight years ago, all these in-use elements (i.e., the assemblage costs) are eight years old on the effective date of the appraisal and should be depreciated, because they are not new and do not have the same remaining useful life as if they were new. Based on an age of eight years and an expected life of 20 years, depreciation is estimated to be 40 percent (8 ÷ 20 × 100 = 40 percent). Applying 40 percent to the cost new of $625 results in a value for the installation and connections of $375. The sum of the base unit ($6000) plus the value of the installation and other assemblage costs ($375) equals the indicated *fair market value in continued use* or *fair market value installed,* that is, $6,375.

The above has been simplified to illustrate the concept behind the sales comparison approach. The numbers used above are, of course, fictitious, but a wide range in selling prices is realistic. To conclude a specific number from that broad a range (as described in the example) requires supporting data and a strong knowledge of the marketplace; the better the data, the better the answer.

Appraising a Group of Assets

There are many reasons for appraising a group of assets such as a production line: to determine fair market value at the end of a lease, for acquisition or sale, or for insurance purposes. Although many of the methods just discussed (concerning the appraisal of a single unit) are applicable to the appraisal of a group of assets, appraising a group of assets introduces complexities not encountered when appraising a single asset. This is especially true when using the sales comparison approach to appraise a group of assets under the premise of *fair market value installed* or *fair market value in continued use.*

The appraiser will need to be familiar with the subject property's industry. Information that will assist in the valuation process can be obtained from trade publications, library research, and interviews with the client and used equipment dealers. The research should include the past, current, and projected economic condition of the industry.

In the following examples, it is assumed that the client has requested value estimates for the following premises of value:

1. Fair Market Value
2. Fair Market Value in Continued Use
3. Fair Market Value Installed—Production Line
4. Fair Market Value Removal—Production Line
5. Orderly Liquidation Value—Production Line
6. Forced Liquidation Value—Production Line
7. Liquidation Value in Place—Production Line

Each value concept will be discussed as it applies to the valuation of a production line or other group of related assets. To illustrate the appraisal methods, the following production line will be used as an example:

Complete Vacuum Molding Line, 1985 ABC Model X, 100' Long × 3 Mold Wide Index
Overall Condition at Time of Inspection: Good

Location:	Pennsylvania
Effective Date of Appraisal:	September 1997

Fair Market Value—Production Line

Assume that the appraiser's research has revealed market sales by reputable used equipment dealers specializing in plastics industry equipment. During the verification process, calls to the used equipment dealers indicate the information already obtained by the appraiser was accurate in all important respects. For purposes of the following examples, assume the appraiser has (1) analyzed the market sales; (2) compared the market sales to the subject; (3) made appropriate adjustments to the comparables (as discussed earlier in this chapter); and (4) concluded that the *fair market value* of the subject is $200,000.

Fair Market Value in Continued Use—Production Line

In determining the *fair market value in continued use by the sales comparison approach* of a group of assets such as a production line, two additional steps are taken to those previously described. First, the appraiser adds to the value of the base unit, the value contribution of the installation, and other assemblage costs required to get the base unit "up and running" and contributing to the overall operation of the production line or facility. Second, the appraiser must address whether there are sufficient business earnings to support the value indication obtained by adding the value of the assemblage costs to the used market value of the base unit at the appropriate level of trade.

Adding the Value Contribution of Assemblage Costs

Appraising assets under the premise of continued use requires adding to the base unit value the value contribution of the costs required to get the base units installed in the buyer's plant and ready to operate. In effect, the appraiser converts the market price of the base unit into the *fair market value installed* (or *fair market value in continued use*) price. In this chapter, the various costs that accomplish this conversion will be referred to as "assemblage costs" (and the value contribution of "assemblage costs" will be referred to as "assemblage value").

Thus, to determine either *fair market value in continued use* or *fair market value installed,* the appraiser replicates the actions of a buyer who desires to assemble an operating package of assets from the used equipment market (at an end-user-to-end-user or used-equipment-dealer-to-end-user level of trade). The appraiser, in effect, "purchases" base units (i.e., individual assets) in the used market, and then adds the value contribution (i.e., assemblage value) of those assemblage costs required to make the base unit an operable one contributing (or capable of contributing) to the overall operation of the facility or production line. Typical assemblage costs include sales tax; costs of dismantling, moving, and setting in place; freight costs necessary to get the assets to the owner's site; rebuilding or retrofitting costs (if necessary); installation costs, including connections, foundations, and millwright work; connection costs, including piping, wiring, and instrumentation; design, engineering, or evaluation costs (if necessary); start-up and testing costs; and any other direct or indirect costs that are normally required to place the asset in service. These are the same costs discussed in the cost approach.

Once these assemblage costs are incurred, the used machine is installed and ready to operate. The sum of the base unit and assemblage costs represents the market price for the machine plus the current cost new to make the machine actually or potentially operable.

Up to this point, depreciation of the assemblage costs has not been considered. In most appraisal situations, assemblage costs should be depreciated.[6] Suppose that ten years ago a new asset was purchased and installed, and now that asset is being appraised. Because the asset has been operating for ten years, both the base unit and the assemblage costs (for freight, installation, connections, etc.) are not new and both have shorter remaining useful lives than they did when new. The fact is that the asset, including its connections and installation, is not new. Assuming that the depreciation affecting the value of the base unit has already been measured by the used equipment market,[7] the appraiser still needs to calculate the loss in value of the assemblage costs by depreciating the replacement cost new of the assemblage costs, using the same techniques described in the cost approach.

To illustrate one method of depreciating assemblage costs, the following additional information is provided for the vacuum molding line example previously discussed:

Description:	**Complete Vacuum Molding Line, 1985 ABC Model X**
Fair Market Value as Determined by Sales Analysis (as discussed above in *fair market value—Production Line*):	$200,000
Replacement Cost-New:	$500,000
Indicated Depreciation from All Causes:	60%[8]

Note: Figures are for illustration only and are not intended to suggest actual market values of this type of equipment.

The appraisal depreciation that is inherent in the equipment itself, which is 60 percent, is already reflected in the $200,000 *fair market value* of the uninstalled production line that was obtained from the used equipment market.[9] Adding the depreciated value of assemblage costs will provide an indication of *fair market value installed* or *in continued use.* The best place to obtain installation and other assemblage cost information is from the engineering department of the company owning the asset in question; if it is not available from that source, the appraiser should estimate these costs.

Assume that after analysis of all assemblage costs pertaining to the vacuum molding line, it is determined that the current replacement cost new of all assemblage costs is $100,000. Assuming that the remaining useful life of the assemblage costs cannot exceed that of the production line itself, it is logical to depreciate the assemblage costs using the same depreciation factor that the used equipment market has, in effect, applied to the production line itself, or 60 percent.

The next step is to quantify the contribution to value of the assemblage costs and add these to the value previously determined for the production line:

Fair Market Value (FMV) of Production Line:	$200,000
Replacement Cost New of Assemblage Costs:	$100,000
Less Depreciation of Assemblage Costs @ 60%:	($ 60,000)[10]
Preliminary Indication of FMV Installed or in Continued Use:	$240,000

Previously it was stated that in determining *fair market value installed* or *in continued use,* two additional steps are added to the steps taken to measure *fair market value*: first, adding the depreciated value of the assemblage costs, and second, determining whether there are sufficient business earnings to support the value conclusion as to the underlying assets. It should be noted that although our discussion of business earnings has been postponed until now to facilitate the presentation of sales comparison approach techniques, in the real world it may be that the business earnings are analyzed before the equipment process begins or is completed.

Are There Sufficient Business Earnings to Support the Value Conclusion?

The definition of *fair market value in continued use* includes an *assumption* that there are sufficient business earnings to support the value conclusion as to the assets in question. The appraiser has two options for dealing with this issue. The first option is to assume that there are sufficient earnings. The second option is to use income approach methods to actually determine whether there are sufficient earnings. If the first option is selected, the appraiser must ensure that the appraisal report clearly states that the value reported for the assets in question *assumes* that business earnings are sufficient to support the value conclusion; otherwise the appraisal report may be misleading.

There may be times, however, when the purpose, use, or other requirements of the appraisal preclude the appraiser from assuming that earnings are sufficient to support the value conclusion. In such cases, the appraiser will need to actually determine whether the business earnings are sufficient to support the value conclusion. The appraiser can personally conduct the analysis or, if not qualified, involve another appraiser.

Using the previous example, suppose that the option of merely assuming sufficient earnings is not available to the appraiser. Recall that we had just concluded that the "preliminary indication"[11] of *fair market value installed* or *in continued use* is $240,000. Suppose that further investigation reveals (1) the vacuum molding line is one of five production lines at the plant; (2) the total facility generates $300,000 of net cash flow annually; (3) the subject line contributes approximately 20 percent of the total net cash flow; (4) the subject production line is expected to generate this cash flow for ten more years; and (5) the discount rate is 12%. Because the present value of $60,000 ($300,000 × 20%) per year for ten years is greater than $240,000,[12] the appraiser can conclude that there is, in fact, sufficient cash flow to support the preliminary indication of $240,000 as the correct *fair market value installed* or *in continued use.*

On the other hand, if the subject line were generating net cash flow of only $30,000 per year, the present value of the cash flow stream generated by the line would be substantially less than the $240,000 preliminary indication of value.[13] This result would suggest the possibility that the value of the subject has been further reduced, from what it would otherwise be, by plant-specific obsolescence that the used equipment market is not capable of measuring. Plant-specific obsolescence refers to a condition within the particular plant that reduces the utility or profitability of the subject property. At this point, appraisers should ask themselves what kinds of obsolescence, especially economic obsolescence, the used equipment market is not capable of measuring.

An example of plant-specific obsolescence not measurable by the used equipment market would be a production bottleneck caused by the slower capacity of equipment installed ahead of the subject in the production process. Assume the following: (1) that the subject production line has a capacity of 1,000 units per hour but that slower capacity equipment installed ahead of the subject limits it to actual production of only 800 units per hour; (2) that there is sufficient economic demand for the extra 200 units; (3) that except for the slower capacity equipment that is creating the bottleneck, there is no reason why the subject line could not produce 1,000 units per hour; and (4) that the company owning the facility plans to expend the capital required to remove the production bottleneck (in appraisal language, to cure the obsolescence), but that due to constraints on the company's capital spending, it will be three years (from the effective date of the appraisal) before the company can cure the obsolescence.

In this instance, the appraiser has several alternatives for determining whether the subject's *fair market value installed* or *in continued use* is affected by obsolescence not measured by the used equipment market. If there are market sales of 800 unit-per-hour vacuum molding lines, those could be used as comparables instead of market sales of 1,000 unit-per-hour lines. There are two possible problems with this analysis: first, the bottleneck is scheduled to be removed in three years; and second, suppose (as is often the case in the real world) that there are no sales of 800-unit-per-hour lines available to be used as comparables (perhaps because, for example, the next lower-capacity model available for sale has a 500-unit-per-hour capacity).

A second possible way to analyze the situation would be to focus on quantifying the reduction in utility (and thus value) caused by the line operating at only 80 percent of its rated capacity (the reader is referred to the discussion of this subject in the cost approach chapter). A possible problem with this analysis is that this inutility is not permanent.

A third possible way of analyzing the situation would be to determine the present value of the lost cash flow caused by the bottleneck during the next three years.[14] The kind of information necessary to make this calculation is often available, if the appraiser is able to identify the condition in the first place. For example, usually the management of a process plant is able to provide the appraiser with the kind of data that enable the appraiser to estimate the present value of lost cash flow caused by a condition such as the above.

The essential point here is that when appraising *fair market value installed* or *in continued use,* the appraiser should remember that the used equipment market may not "automatically" measure all depreciation and obsolescence. Appraisers should always ask themselves whether there is additional depreciation or obsolescence that the used equipment market does not reflect. The appraiser may not become aware of the possible existence of unmeasured obsolescence unless the *preliminary* value indication of the sales comparison approach is independently checked against income- or cost-based analyses.

Fair Market Value Installed—Production Line

It can be seen from the above discussion that there is nothing a MTS appraiser adds to the costs (value) to reflect a *fair market value in continued use* versus *fair market value installed.* Except for sale and leaseback purposes, the appraiser will rarely be called upon to perform a *fair market value installed* appraisal of only one production line in a facility consisting of multiple lines. If the facility were sold, it would probably be sold as an overall facility and it would be necessary to value all of the lines. However, if only one line needed to be valued, such as in an appraisal in connection with a sale and leaseback, it should generally consider the same factors as a *fair market value in continued use* appraisal, except that it would not be necessary to assume (or independently determine) that there are sufficient earnings to support the underlying asset value conclusion. Other things to consider would be the segregation of utilities, workforce, and other considerations that would, generally, point to a lower value than *fair market value in continued use.*

Fair Market Value Removal—Production Line

This valuation concept is similar to the *fair market value* concept, except that the cost of removal must be considered. Recalling that the market comparables (in an earlier example) indicated a *fair market value* of $200,000 for the subject vacuum line. This means that a potential buyer would pay no more than $200,000 for the equipment if it were ready to be picked up and delivered to the buyer. However, if the equipment is still installed in a facility, the prudent buyer will not pay the full $200,000 because he or she will still have to incur the removal costs of dismantling, crating, and shipping the equipment. Assuming the removal cost for the vacuum molding line is $30,000, the *fair market value – removal* would be $170,000 ($200,000 – $30,000). The appraiser must be aware, however, that buyers of installed equipment under liquidation or market value conditions often make conscious (and sometimes unconscious) adjustments to the price offered to cover any removal or relocation costs.

Orderly Liquidation Value—Production Line

Orderly liquidation value may be very close to *fair market value – removal,* the difference being that under the premise of orderly liquidation there is a limited period in which to sell. The seller is compelled to sell, although there is not the same sense of immediacy or urgency that is assumed in a forced liquidation sale (see *forced liquidation value* discussion below). The orderly liquidation sale may be necessitated by a bankruptcy court ruling, or by a leasing company, a bank, or other institution holding a note. The company may need to sell assets for some other reason.

Forced Liquidation Value—Production Line

Under the *forced liquidation value* premise, a sense of immediacy or urgency affects the period of time and circumstances of the sale. In *forced* liquidation, the means of sale are a properly advertised auction or other public sale. It is important to note that in some instances, auction sales may result in prices equal to *fair market value* because of the desirability of the equipment, or proper advertisement of and high attendance at the auction. At other times, auction sales can result in prices substantially below the assets' *fair market value*. It is incorrect to automatically assume that the results of an auction always produce a *forced liquidation value*. An independent study of the circumstances of each auction sale should be conducted. An auction is a method of sale, not a value premise. The auction attendees have a wide range of motivations that are reflected in their bids, which may or may not mirror the premise of forced liquidation value.

Liquidation Value in Place—Production Line

Under the *liquidation value in place* premise, because of the limited time to complete the sale, the value for the subject production line would usually decrease considerably. The dilemma the appraiser faces is how to quantify the amount of this decrease. In analyzing the market data gathered, the appraiser should always ask how long were the similar items (about which market data has been collected) exposed to the market and was there an excess of this type of asset on the used equipment market at the time of the sale.

Appraising an Industrial Facility

As with the appraisal of a group of assets, applying the sales comparison approach to an entire industrial facility introduces complexities not encountered when appraising a single asset. This is especially true when using the approach to appraise an operating facility under the premise of *fair market value in continued use* or *fair market value installed* (capable of being used).

The appraiser using the sales comparison approach to determine the *fair market value in continued use* of an entire industrial facility has, in theory, two alternative methodologies. The first alternative is to use the methods already discussed to, in effect, replicate the process of a plant owner assembling an operating package of assets from the used equipment market, that is, purchase individual assets in the used market; dismantle and move them; and add the value of the various assemblage costs, such as tax, freight, rebuilding, installation, connection, and other costs, including any adjustments for level of trade.

The second alternative is to compare the subject property (an entire industrial facility) to sales of comparable industrial facilities. In the following discussion, these sales will be referred to as "whole plant sales," and the method will be referred to as "the whole plant sales comparison approach."

Limitations of the "Whole Plant" Approach

Before discussing the whole plant sales comparison approach, it is important to note that this approach will often not be feasible. Most industrial facilities infrequently change ownership. Their unique characteristics make comparability difficult to evaluate and adjustments difficult to quantify without undue speculation. Most industrial facilities change ownership based on the parties' perceptions of the present value of the future cash flow that the facility is capable of generating. Differences in perceived profitability are crucial and often explain the large difference in the selling prices of ostensibly comparable facilities. The details of industrial facility transactions are usually confidential. Often even the selling price, let alone other vital details such as future cash flow projections, is difficult or impossible to obtain.

In addition to the above problems, the reported selling price of an industrial facility often includes the value of working capital, other business enterprise intangibles, or other assets (such as land) that are not included in the subject property the appraiser needs to value. The value of these assets is often difficult to segregate from the total selling price. For example, the total selling price of a paper mill may include the value of large timber reserves, in addition to working capital and other business enterprise intangibles.

Notwithstanding the above, there are situations in which these difficulties do not exist or can be overcome, and in those cases the whole plant sales comparison approach may be useful. The approach is best used when the comparables are virtually identical to the subject and have been sold relatively recently and in an active market. Obviously, the use of this approach requires extensive research and analysis of comparability, adjustments, and other issues. To employ this approach, it may be that the MTS appraiser will need to have an adequate understanding of business or real property valuation, or to work closely with appraisers from those disciplines.

Methodology

The whole plant sales comparison approach compares the subject property to whole plant sales of comparable industrial facilities. It is based upon the observation of actual market sales involving comparable facilities. The market sales that are initially identified as possible comparables should be carefully analyzed to ensure that their profitability and risk characteristics are sufficiently similar to the subject facility's. While the sold facilities may differ from the subject facility, the effect of these differences should be minimized by initially selecting transactions that possess characteristics as similar as possible to the subject facility.

After gathering market sales data, the appraiser makes adjustments to the comparables to reflect differences between them and the subject with respect to numerous factors, including

- current profitability,
- future profitability,
- future growth in profitability,
- product,
- future market for the product,
- transaction date,
- transaction terms,
- market conditions,
- physical characteristics such as facility size and capacity,
- physical condition,
- deferred maintenance,
- raw material costs,
- labor costs,

- labor union impact,
- facility location,
- distance from market for finished goods, and
- other factors affecting value.

After the selling prices of the comparables have been adjusted to equate them with the subject, the adjusted selling prices are then converted to a common unit of measure such as price per unit of output capacity. An indication of the subject facility's value can then be developed by multiplying its productive capacity by the price per unit of output derived from the comparable facility transactions.

Example 2: Whole plant sales comparison approach.

You have been asked to determine the *fair market value in continued use* or *fair market value installed* of a 500-megawatt coal-fired electric generating facility.[15] You have performed an income approach (or had another appraiser do it) and are now considering whether a whole plant sales comparison approach is feasible. Industry research and discussions with knowledgeable persons have disclosed a significant number of relatively recent sales (say, within the past two years) of coal-fired electric generating facilities. You are aware that the whole plant sales comparison approach is best applied when the comparables are virtually identical to, or at least extremely similar to, the subject property. Potential sources of information regarding electric generating facility transactions include Securities and Exchange Commission (SEC) filings, corporate annual reports, industry trade publications, and news articles.

After much research, you have concluded that all but six of the relatively recent market sales do not meet the threshold requirements for comparability. Some of the sales have involved facilities using other sources than coal—gas, oil, or hydroelectric—to produce power. You have rejected other sales because of significant differences between them and the subject with respect to future cash flow potential, growth in cash flow, and capacity. You are now satisfied that the six sales that have survived the "comparable qualification" process are extremely similar to the subject property; that is, they are truly comparable to the subject. These six sales are listed in the table below.

Purchaser	Selling Price	Capacity (kW)	$ / kW
Sale 1	$ 425,000,000	550,000	$773
Sale 2	350,000,000	425,000	824
Sale 3	400,000,000	525,000	762
Sale 4	365,000,000	450,000	811
Sale 5	400,000,000	500,000	800
Sale 6	310,000,000	400,000	775
Subject	Unknown	500,000	Unknown

Table 4.2. Comparable Sales Electric Generating Facilities

Analysis of these transactions indicates the selling price per kilowatt ranges from $762 to $824 and averages approximately $790. Because you eliminated all market sales except those that were virtually identical to the subject in all important respects relating to value, you conclude that no adjustments to the comparables are needed. The indicated range of values for the subject is $380 million to $412 million ($762/kW × 500,000 kW = $381 million and $824/kW × 500,000 kW = $412 million). The best use of the whole plant sales comparison approach may be to indicate a reasonable range of value rather than a specific point value estimate, in which case the appraiser could stop here.

If a point value estimate is required, assume that sale 5 is virtually a direct match to the subject, in which case the appraiser might give it extra weight and conclude that the subject's *fair market value in continued use* (or *fair market value installed)* is \$800 per kilowatt, or \$400 million (\$800/kW × 500,000 kW = \$400 million). Once the \$400 million conclusion of value is reached by the sales comparison approach, it should be compared to the results indicated by the income approach (which we have assumed was also done). If the results of the sales comparison approach are substantially higher than those of income approach, there may be additional plant-specific functional or economic obsolescences that have not been measured by the sales comparison approach.

Example 3: Whole plant sales comparison approach using percentage of cost technique.

Your task is to estimate the *fair market value in continued use* or *fair market value installed* of a coal preparation plant. The subject is ten years old and rated at 800 tons per day. The facility receives maintenance on a preventive basis and the plant is in reasonably good condition. Investigation reveals a recent upturn in the coal mining industry, and because of this, there seems to be an increased demand for coal preparation plants. Having already concluded a value by the cost approach (and possibly the income approach), your investigation of the facts has also disclosed two sales of coal preparation plants within the past year.

The first sale involved the purchase of an operating company, including both its tangible and intangible assets. The assets of the company included the coal preparation plant, a fleet of trucks, other aboveground mining equipment, and two long-term contracts negotiated before the sale of the company. The company's primary business is to process coal for two nearby strip mines. You have been told that the operating company was purchased for approximately $10 million. You talk to the new owner, who says the $10 million is "a little high" but does not disclose the exact purchase price; he adds that he "wanted the contracts."

Based on the above information, you conclude that the first sale is not a valid comparable to the subject property because this was a sale of an operating company that included intangible assets, including contracts and other business enterprise intangibles (such as net working capital). In addition, you have not been able to obtain sufficient verified facts to serve as a basis of comparison.

The second sale involved the purchase of a coal preparation plant rated at 1,000 tons per day. The facility had been idle for two years before the sale. Its former owner declared bankruptcy in the midst of a recession. The bank retained ownership and contracted with a local equipment broker to maintain the facility and subsequently to sell it in operating condition. It was recently purchased by a large coal company that relocated it so it could operate in conjunction with another of the purchaser's nearby operations (note that although the plant was idle for two years, the purchaser clearly purchased it with the intention of operating it). The dealer who sold the plant confirmed that the bank had held the asset, hoping the market would rebound, which in fact it did. The dealer also said there was a substantial interest from many companies, but the facility was ultimately purchased by a company that had another operation nearby.

The second-sale plant was eight years old and required minimal capital expenditures since the broker operated the plant periodically to ensure that everything worked. The buyer tells you that the purchase price was \$3.5 million and that an additional \$1.5 million was spent to dismantle, move, and install it at their nearby site. The buyer is satisfied and feels the purchase was a good deal, since the current replacement cost new of a plant of the same capacity is approximately \$12 million installed. The buyer is planning to operate the purchased plant at a mine with a limited remaining economic life, estimated to be five years, and the purchase of a new plant for that application could not be justified.

Since you rejected using the first sale, you are left with only this one market sale of a coal preparation plant, and you must decide if this second sale can be used. Although the plant was idle for two years before the sale, the buyer clearly purchased it with the intention of operating it. You have confirmed with the dealer who sold the plant that the bank held the asset, hoping that the market would rebound, which in fact it did. You have also learned several buyers were interested in buying the plant for continued use. On the basis of these facts, you decide this is a valid comparable sale. In this situation, it would be appropriate to use the percentage-of-cost technique (discussed earlier). The ratio of the \$5 million purchase price (including relocation cost) to the current cost new of \$12 million is approximately 40 percent (\$5 million ÷ \$12 million = 0.417 or 41.7 percent, rounded to 40 percent).

The subject is somewhat less desirable than the comparable because it is a little smaller and two years older, and the subject's installation and other "assemblage" value components are not new. Thus, although the subject is very similar to the comparable, because it is somewhat less desirable than the comparable, something less than 40 percent of cost new would be appropriate for the subject. For illustration purposes assume the conclusion that 35 percent is a reasonable ratio of value to cost new for the subject. If the cost new of the subject is determined to be $10 million (it is smaller than the comparable, and so the cost new would be less than $12 million), then the subject's *fair market value in continued use* (and *fair market value installed)* is approximately $3.5 million ($10 million × 35%).

Once the $3.5 million conclusion of value is reached by the sales comparison approach, it should be compared to the results indicated by the cost approach (which we assume was also done). If the results of the cost approach are substantially higher than $3.5 million, the sales comparison approach may be used as a basis to conclude the existence of additional depreciation (probably economic obsolescence). If the results of the two approaches are reasonably close, both value conclusions are probably reasonable. If the results of the cost approach are significantly lower than $3.5 million, the value derived by the sales comparison approach may be a more reasonable answer, reflecting additional desirability in the marketplace.[16]

Key Points

- The MTS appraiser uses the sales comparison approach to indicate value by analyzing recent sales (or offering prices) of property that is similar (i.e., comparable) to the subject property. If the comparables are not exactly like the assets being appraised, adjustments are made to the selling prices of the comparables to equate them to the characteristics of the assets being appraised. The basic procedure is to gather data on sales and offerings of similar properties, determine their comparability to the subject property, determine the appropriate units of comparison, collect and array the data, analyze and adjust the data, and apply the results to the subject.
- The sales comparison approach is most reliable when there is an active market providing a sufficient number of sales of comparable property that can be independently verified through reliable sources. Examples of assets generally having markets meeting this definition are automobiles and trucks, computers, aircraft, and other assets with an identifiable market.
- The approach is not feasible when the subject property is unique. Even if the subject property is not unique, the approach will generally not be feasible if an active market does not exist. An inactive market or limited number of sales of comparable property often indicates a lack of demand and the existence of economic obsolescence, which may be measured using the income or cost approaches.
- The implementation of the sales comparison approach may differ significantly depending on whether the subject is an individual asset, a group of assets, or an entire facility.
- The appraiser should remember that adjustments are made to the comparables, not the subject property. Elements of comparison are the comparable's vintage, effective age, condition, capacity, features (accessories), location, manufacturer, motivation of parties, price, quality, quantity, time of sale, and type of sale.
- The three most commonly used techniques for establishing value of individual items of machinery and equipment using the sales comparison approach are direct match, comparable match, and percentage of cost.

- Using the sales comparison approach to appraise a group of related assets or an entire industrial facility introduces complexities not encountered when appraising a single asset. This is especially true when using the approach to appraise a group of assets under the premise of *fair market value in continued use* or *fair market value installed* (capable of being used).
- Appraising machinery and equipment under the premise of continued use requires adding to the value of the base units the value contribution of the costs required to get the base units installed in the buyer's plant and ready to operate. In effect, the appraiser converts the market price of the base unit into *fair market value in continued use* or *fair market value installed* (capable of being used).
- The definition of *fair market value in continued use* includes an *assumption* that there are sufficient business earnings to support the value conclusion as to the assets in question. The appraiser has two options for dealing with this issue. The first option is to assume, without verification, that there are sufficient earnings. The second option is to use income approach methods to actually determine whether there are sufficient earnings.
- There are a limited number of situations in which it may be feasible to use the "whole plant" sales comparison approach to value an entire industrial plant. This approach will often not be feasible for the following reasons: (1) most industrial facilities infrequently change ownership; (2) their unique characteristics make comparability difficult to evaluate and adjustments difficult to quantify without undue speculation; (3) most facilities change ownership based on the parties' perceptions of the present value of the future cash flow that the facility is capable of generating; differences in profitability are crucial and often explain the large difference in the selling prices of ostensibly comparable facilities; (4) the reported selling price of an industrial facility often includes the value of net working capital, other business enterprise intangibles, or other assets (such as land) that are not included in the subject property the appraiser needs to value.
- Notwithstanding the above, there are situations in which these difficulties do not exist or can be overcome, and in those cases the whole plant sales comparison approach may be useful. The

approach is best used when the comparables are virtually identical to the subject and have sold relatively recently in an active market. The use of this approach requires extensive research and analysis of comparability, adjustments, and other issues. Just as real property or business appraisers may need to have an adequate understanding of MTS appraisal techniques, it may be that the MTS appraiser will need to have an adequate understanding of business or real property valuation, or work closely with appraisers from those disciplines.

Notes

Note: Portions of this chapter have been repeated, with author's assent, from "Fair Market Value Concepts," a chapter authored by Robert S. Svoboda, ASA, in the book Appraising Machinery and Equipment. Herndon, Va.: American Society of Appraisers, 1989.

[1] The sales comparison approach is sometimes referred to as the "market approach" or "market data approach."

[2] The sales comparison approach does not measure the kind of obsolescence that the used market could not possibly reflect, such as functional or economic obsolescence caused by (1) the relationship of the subject machine to other machines in its production line (e.g., a production bottleneck ahead of the subject property); (2) the subject's relationship to a building or other structure in which it is located (e.g., lower productivity or other obsolescence caused by multiple buildings or poor layout); (3) localized economic obsolescence due to regional raw material shortages; or (4) other conditions or circumstances that the used market could not possibly reflect. Appraisers should always ask themselves what kinds of obsolescence the used equipment market cannot measure.

[3] The income approach may be the best method for measuring economic obsolescence but some forms of economic obsolescence can be at least partially measured by the cost or sales comparison approaches; see discussion of economic obsolescence in the cost approach chapter.

[4] The definitions are given in chapter 1.

[5] It should be noted that sales from outside the subject's market area may require adjustment in order to be made comparable.

[6] The issue of whether assemblage costs should be depreciated is not settled at this time. The M&TS Committee of the American Society of Appraisers currently takes the position that in most appraisals assemblage costs should be depreciated. However, there may be occasions when it is appropriate not to depreciate assemblage costs. For a more detailed discussion of this issue, see "Valuing the 'Installed' in FMV-Installed," Frank Hanrahan, *The M&TS Journal*, vol. 11, no. 2, pp. 8–11 (Fall 1994) and "Fair Market Value in Continued Use vs. Installed," Robert Podwalny, ASA, *The M/TV Journal*, vol. 10, no. 2, pp. (Fall 1993).

[7] But see footnote 2.

[8] 1 – ($200,000 / $500,000) = 60%

[9] But see footnote 2 and further discussion below.

[10] Thus the value contribution of the installation is $40,000.

[11] Preliminary because we had not yet dealt with the adequacy of earnings issue.

[12] The present value annuity factor for ten years at a discount rate of 12% is 5.650223; $60,000 × 5.650223 = $339,013.

[13] The present value annuity factor would be the same as the previous calculation but the present value would be only $169,507 ($30,000 × 5.650223 = $169,507).

[14] It is important to note that any obsolescence penalties quantified by sales comparison, income or cost approach methods cannot be greater than the cost to cure the problem.

[15] The authors do not intend to imply that electrical generating facilities, as a class of property, necessarily meet the requirements discussed above for use of this approach; the selection of this particular kind of property is for purposes of illustration only.

[16] Remember that the results of the cost and sales comparison approaches must also be considered in light of the results of the income approach.

5

Income Approach

Objectives:

1 Provide an understanding of the relevance of the income approach to the cost and sales comparison approaches.

2. Provide a basic understanding of income approach theory.

3. Explain same appropriate applications for the income approach.

4. Introduce the basic terminology relevant to the income approach.

5. Outline the basic steps to perform an appraisal using the income approach.

6. Demonstrate application of the income approach through an example.

The value of a property can be estimated by its expected future benefit to its owner. This is a widely accepted concept both within the valuation community and among those using valuations. The income approach to value is not widely used today by most MTS appraisers; the reasons given include the difficulty in determining income that can be directly related to a specific asset, the concern over the reliability of income forecasts, and the multitude of variables involved in this valuation approach.

While it is true that the MTS appraiser faces many challenges in applying an income approach to value—and there are many instances in which this approach is not feasible—it is also true that when properly applied, the income approach is a powerful valuation tool that can enhance both the quality and credibility of the valuation, especially when used in conjunction with other valuation approaches. Since USPAP specifies that an income approach to value, along with its degree of applicability to the specific situation, be considered in every valuation, the professional MTS appraiser must learn about this approach and its proper application.

The specific goals of this chapter are to help the MTS appraiser understand the relevance of the income approach to the cost and sales comparison approaches; to provide a basic understanding of income approach theory; to help the appraiser recognize the appropriate and common applications of the income approach; to outline the strengths and weaknesses of the income approach to value; to introduce the basic terminology relevant to the income approach to value; to outline the basic steps to perform an income approach on specific assets or groups of assets; to describe the basic steps used to develop discount and capitalization rates; and to demonstrate, by example, an application of an income analysis applied to the machinery and equipment of a business. After completing this chapter, the reader will understand that the income approach may be a practical tool in the valuation of machinery and equipment.

Theory

Present Value of Future Benefits

Before addressing specific methodologies for applying the income approach in an MTS valuation, a few important points that underlie this approach must be understood. The decision to purchase any business asset, whether a piece of land, a building, a business, an automobile, or an item of machinery and equipment, is typically an investment decision. Investment decisions are made based on the expected return to be generated by the investment, the period of time over which the return will be earned, and the risk of not receiving the expected return. Investments are typically made in one of two forms: equity or debt. In an equity investment, the primary party purchases a property or an interest in a property. With a debt investment, money is lent to a second party for the purchase of a property or an interest in a property. Before an investment decision is made, the investor must understand all of the future benefits from making the investment. The range of benefits is almost limitless; some of the more common ones include interest, dividends, capital appreciation, business synergies, and tax incentives. Since these benefits typically accrue to the investor over time, an investor needs to know the current value of all the identifiable future benefits. An investment decision is based on the present value of the future benefits to be earned by the investment, and the value of a particular asset is represented by the present value of its expected future benefits. This is the foundation for the income approach to value as it applies to MTS assets. The income approach is a method for measuring the present value of the asset's expected future benefits.

Time Value of Money

One of the essential concepts in the income approach is the *time value of money*. The time value of money concept encompasses several important terms including present value, future value, compounding, and discounting. Stated simply, the time value of money concept is that a dollar received today is worth more than a dollar to be received in the future, because the dollar received today can be invested and earn interest or other benefits. Stated another way, a dollar is worth more today because it can be used to buy goods and services (i.e., it is liquid) and there is no risk associated with waiting.

Present value represents the value today of something that will be received in the future. For example, if you were going to receive one dollar five years from now, what would that future payment be worth if you received it today instead? The answer is that it would be worth the amount of money that would need to be invested today in order to have one dollar five years from now, at a specified interest rate or rate of return. The process used to determine the present value is referred to as *discounting.*

On the other hand, *future value* is the value in the future of something that is owned today. For example, if you invested one dollar today, and received a 12 percent rate of return for five years, the future value would be about $1.76 at the end of five years. The process of determining the future value is referred to as *compounding.*

In addition to determining the present or future value of a "one-time" benefit, the present or future value of "multiple" benefits can also be determined. A series of equal payments over time is called an *annuity*. The present value of an annuity is the value of a series of future benefits received today. The reverse—future value of an annuity—is the value of a series of future benefits at a future point in time. Various generally accepted mathematical formulas for determining the present and future values of both one-time benefits and annuities are shown in table 5.1.

Rate of Return, Discount Rate, and Capitalization Rate

Three essential definitions involved in the income approach are *rate of return, discount rate,* and *capitalization rate.* The *rate of return* concept is based on the assumption that virtually every investment, including machinery and equipment, is made with the expectation of receiving a return from that investment. For example, a stock investor expects returns in the form of dividend income and stock price appreciation, above and beyond the cost of the original investment. Likewise, the expectation of a property owner investing money into materials to build an apartment house would be the hope of generating a return in the form of rental income or property appreciation. Alternatively, machinery and equipment could be purchased to start a manufacturing company with the hope of generating income. Regardless of the investment—money, time, resources—the investor expects a certain return from that investment.

1. **Future Value**—A dollar in hand today is worth more than a dollar to be received next year because today it could be invested and earn interest. The process of finding future values is called compounding. The future value, FV, at the end of "n" periods is the present value times 1 plus the interest rate raised to the "n" power.

PV = present value.
i = interest rate.
n = number of periods.

$$FV_n = PV \times (1+i)^n$$

2. **Present Value**—The present value of a sum due in "n" years in the future is the amount which, if it were on hand today, would grow to equal the future sum. The process of finding present values is called discounting. In words, the present value, PV, is the future value at the end of "n" periods, times the quotient of 1 divided by 1 plus the interest rate, raised to the "n" power.

FV_n = future value at the end of "n" periods.
PV = present value.
i = interest rate.
n = number of periods.

$$PV = FV_n \times \frac{1}{(1+i)^n}$$

3. **Future Value of an Annuity**—An annuity is defined as a series of payments of an equal, or constant, amount of money at fixed intervals for a specified number of periods. When the future values of each of the payments are summed, their total is the future value of the annuity.

FVA_n = future value of an annuity for "n" periods.
PMT = periodic payment.
i = interest rate.
n = number of payments.

$$FVA_n = PMT \times \left(\frac{(1+i)^n - 1}{i} \right)$$

4. **Present Value of an Annuity**—The present value of an annuity is the lump sum amount of money that, if invested at a specified rate for a specified period of time, would generate a certain payment stream over that period of time.

PVA_n = future value of an annuity for "n" periods.
PMT = periodic payment.
i = interest rate.
n = number of payments.

$$PVA_n = PMT \times \left(\frac{1}{i} - \frac{1}{i(1+i)^n} \right)$$

Table 5.1 Time Value Formulas

Investments are generally made with two basic goals in mind: achieving the return "of" the original investment—getting the original investment back—and receiving a return "on" the investment—getting more than the original investment back. There are virtually unlimited investment alternatives, each with varying degrees of risk. It is important in the investment decision process to make investments that will provide a rate of return that is commensurate with the investment risk. It is this return, in the form of income or cash flow generated by the asset, that is quantified and discounted to estimate value using the income approach.

The *discount rate* is the rate of return used in discounting the future benefits generated by an asset or group of assets, to determine present value. When we use the required rate of return to discount future benefits, we call it a discount rate.

The purpose and context of the valuation will have a great deal to do with the specific methodology chosen to determine the discount rate. In the following case, assume that you are valuing an entire industrial plant that will continue to operate in its current capacity and configuration. Also, remember that the goal in determining any discount rate is to develop a rate that appropriately compensates an investor for the relative risks inherent in investing in the subject versus investing in other alternatives. How the rate is determined is not as significant as its reflection of the level of risk associated with the investment.

It is important for the appraiser to distinguish between the *discount rate* and the *capitalization rate*. Shannon Pratt explains it as follows:[1]

> In the process called discounting, we project *all* expected returns from the subject investment to the respective class or classes of capital over the life of the investment. Thus, the percentage return that we call the discount rate represents the total compound rate of return that an investor in that class of investment expects to achieve over the life of the investment.

> There is a related process for estimating present value, which we call *capitalizing*. In capitalizing, instead of projecting all future returns on the investment to the respective classes of capital, we focus on the return of just one single period, usually the return expected in the first year immediately following the valuation date. We then divide that single number by a divisor called the *capitalization rate*. This process is called *capitalizing*.
>
> As we will see, the process of capitalizing is just a shorthand form of discounting, and the capitalization rate is actually a derivative of the discount rate. That is, the capitalization rate, as used in the income approach to valuation or project selection, is formed by derivation from the discount rate. ...Assuming stable long-term growth in the cash flows available to the investment being valued, the capitalization rate equals the discount rate minus the expected long-term growth rate.

For example, if the discount rate is 12% and expected long-term growth is 2% per year, the capitalization rate is 10% (12% – 2%). The expected long-term growth rate is not just inflation, although inflation is considered in the forecast.

Weighted Average Cost of Capital

One way to determine a discount rate is to consider the rate of return that is required by a potential investor in the asset (in this case, the industrial plant). To accomplish this, the *weighted average cost of capital* can be used. In the following discussion, refer to figure 5.1.

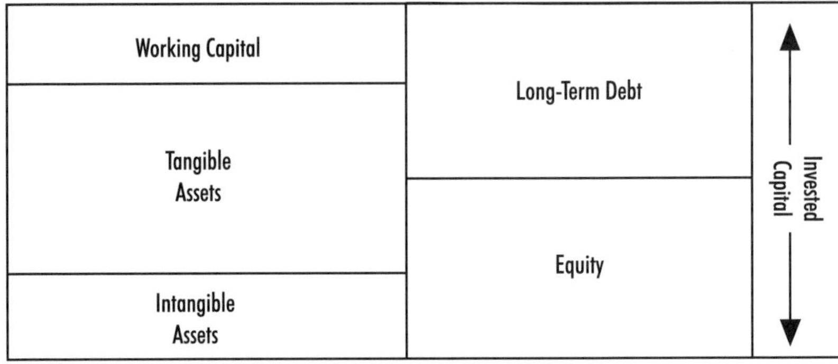

Figure 5.1. Economic balance sheet.

Invested Capital	=	The sum of the net working capital, tangible and intangible assets, or
	=	The sum of long-term debt and equity, or
	=	The business enterprise value.
Working Capital	=	The necessary amount of the current assets needed to operate the businessduring the projected period. Working capital is defined as current assets (CA) minus current liabilities (CL).
Tangible Assets	=	Land, land improvements, buildings, machinery and equipment, office furniture and equipment, and other tangible assets.
Intangible Assets	=	Going concern value, goodwill, and other intangibles such as contracts, assembled workforce, and computer software.

The investments in a business are typically made in the form of debt and equity. Together, they form the invested capital of the business (right side of figure 5.1). The business then takes that capital investment and buys assets (the left side of the diagram). By determining the costs of the capital investment in the business—the rates of return required by the investors who are providing that capital—an indication of the relative risk involved with the subject asset is obtained. The *weighted average cost of capital (WACC)* is the appropriate *discount rate* when valuing the assets of a business by the income approach.

The WACC is an important concept to an investor. An investor's cost of capital is the return that must be provided to an investor for an additional capital investment, such as short-term debt, long-term debt, or preferred and common stock (equity). It can also be referred to as the weighted rate of return that investors require on a company's assets less its non-interest-bearing current liabilities. The cost of a business's capital is directly related to the investor's perceived risk of investing in that business. In other words, the greater the risk that investors perceive in investing in a business, the greater the cost of that business's capital. Therefore, the investor does not consider the components of the cost of capital (debt and equity) to be at equal risk.

The WACC represents the after-tax return on the elements of invested capital weighted by their relative percentage of the capital structure. The relative percentage of these elements in the capital structure should be based on market comparable data. The following table illustrates the computation.

	1 Capital Structure	2 Pretax Cost	3 Tax Effect*	4 After-Tax Cost	5 (1 x 4)
Debt	40%	8.4%	60.0%	5.04%	2.02%
Equity	60%	17.1%	100.0%	17.10%	10.26%
WACC					12.28%

* The tax rate is assumed to be 40%

Table 5.2. Weighted Average Cost of Capital

In table 5.2, the *weighted average cost of capital* represents the after-tax return on the elements of the invested capital of the business, weighted by their relative percentage of the total invested capital. Note that capital structure is based on market values of debt and equity, not as reported in financial statements. An important point is that the returns represent "marginal returns" to the various forms of capital—returns required on the next dollar invested in the business, not on the last dollar invested. Therefore, as the business buys assets that have more risk attached, the required rate of return on the capital being invested in the business will increase. In addition, it is very important that any asset that is purchased by a business generate a rate of return that is at least equal to its cost of capital. If not, the business should probably not buy that asset. Realize, however, that some assets generate greater returns than do others. Finally, in table 5.2, the WACC is made up of the cost of the equity and the cost of debt. As shown later in the chapter, the WACC or discount rate will be applied to a debt-free after-tax net cash flow stream from operations. In other words, there are no deductions in the income (cash flow) projection for interest expense on debt, debt repayment, or interest returns on investments. The cost of equity concept is discussed below.

Cost of Equity

Two of the basic methods that can be used to determine the cost of equity are

- the build-up or summation method, and
- the capital asset pricing model (CAPM) method.

Build-Up Method

The basic theory of the build-up method (sometimes referred to as the "summation method") is that it is possible to identify and quantify the various risk elements associated with an equity investment. Once the risk elements have been identified and quantified, they are simply added together to determine the rate of return required on the investment. Two models can be used to derive the investor's required return on an equity investment. The following table illustrates build-up method #1.

Data Inputs:

1.	Risk-Free Rate (R_f)	5.9%
2.	Economic Risk Premium (E_r)	5.0%
3.	Illiquidity Risk Premium (I_r)	3.4%
4.	Business Risk (B_r)	2.7%
5.	Industry Risk Premium (D_r)	0.8%
6.	General Economic Risk Premium (G_r)	1.2%

The cost of equity is calculated by starting with the risk-free rate (R_f) and adding relevant "risk premiums":

$$\text{Cost of Equity} = R_f + E_r + I_r + D_r + G_r$$

$$\text{Cost of Equity} = 0.059 + 0.050 + 0.034 + 0.027 + 0.008 + 0.012$$

$$\text{Cost of Equity} = 0.19 \text{ or } 19\%$$

Using this method, the required rate of return (cost of equity) is 0.19 or 19%.

Table 5.3. Build-Up Method #1

In table 5.3, the basic concept is to start with a risk-free rate of return and add the incremental risk premium(s) that reflect the risks inherent in the subject equity investment. Investments that are generally considered to be risk-free are U.S. Treasury bills or U.S. Treasury bonds. In table 5.3, the rate on the 20-year Treasury bond was used. The longer-term Treasury bond rate is preferred by most appraisers, since typical equipment investments are not short-term. The quantification of the incremental risk premiums can be difficult. An essential point to remember is that the build-up method is not a rigid formula. Judgment must be used on a case-by-case basis, to assess the risk components that need to be quantified.

A second method used to estimate the cost of equity using the build-up method involves adding a market-based risk premium to the cost of debt. The next table illustrates build-up method #2.

Data Inputs:		
1.	Cost of long-term debt (R_d)	8.0%
2.	Debt risk premium (R_{p2})	6.5%
3.	Additional risk premium (R_u)	5.0%

The cost of equity is calculated by starting with the cost of debt (R_d) and adding the debt risk premium (R_{p2}) and the relevant additional risk premium (R_u):

Cost of equity $= R_d + R_{p2} + R_u$

Cost of equity $= 0.080 + 0.065 + 0.050 = 0.195$

Using this method, the required rate of return (cost of equity) is 19.5%.

Table 5.4. Build-Up Method #2

This method is simple and straightforward. The debt risk premium (R_{p2}) can be derived from data published by Ibbotson Associates in *Stocks, Bonds, Bills and Inflation: Yearbook*. The cost of debt (R_d) is developed from the cost of industrial bonds, and the additional risk premium (R_u) is added to reflect a small-company risk, a specific-company risk, an illiquidity premium, or the inherent risks of a single plant. The advantage of this method is that the cost of debt and the debt premium are both derived from a published source. The only judgment required of the appraiser is in quantifying the additional risk factor. Again, judgment must be used on a case-by-case basis.

Capital Asset Pricing Model Method

The capital asset pricing model method stems from one of the more widely discussed and used developments in modern financial theory: the CAPM. The CAPM was originally formulated by William F. Sharpe, Harry M. Markowitz, and James Tobin to measure the cost of equity capital. Current discussions of its use can be found in various valuation and finance texts, including *Cost of Capital: Estimation and Applications,* by Shannon Pratt.[2]

This model is based on the fact that investors require a rate of return, above a risk-free rate, as compensation for bearing the risks inherent in their investments. In simple terms, the CAPM technique suggests that the rate of return on an asset is a function of some risk-free rate of return plus a risk "premium" which is a function of the amount of risk associated with the asset or investment. The risk premium is determined as a function of the volatility of the asset's price in the marketplace over time, relative to the volatility of prices in the marketplace as a whole over the same period.

The CAPM method, like build-up method #1, uses the risk-free rate of return. This method also uses the concept of "beta." Beta can be defined as the measure of the volatility of the subject investment's return relative to the volatility of returns in the marketplace as a whole. The beta for an individual equity investment, for instance, reflects its risk relative to that of equity investments in general. The beta can be computed, or in many instances has already been computed and published in various reports.

For example, if the price of an individual stock in the marketplace increases 10 percent while the prices of all stocks in the marketplace increase 10 percent, and then the price of that individual stock decreases 10 percent while the prices of all stocks decrease by 10 percent, the volatility of that individual stock would equal the volatility of the market. The stock's beta would then be equal to 1.0. On the other hand, if the price of a stock increases 20 percent while the prices of all stocks in the market increase 10 percent, and the stock then decreases 20 percent when all stock market prices decrease 10 percent, then that individual stock would have twice the volatility of the market as a whole. The individual stock would then have a beta of 2.0. Alternatively, if the stock price increases 5 percent while stock prices in the market increase 10 percent, and then decreases 5 percent while stock prices in the market decrease 10 percent, the volatility of the stock would be half that of the market as a whole. This individual stock, then, would have a beta of 0.5. The higher the beta, the more potentially volatile the stock's price.

The next component of the CAPM method is the "equity risk premium." The equity risk premium is generally determined by looking at the historical mean of the difference between the risk-free rate and the equity market rate of return (published sources of this information are available through Ibbotson Associates).

The last component is the "additional risk premium." The additional risk premium represents unsystematic risk in the specific property being appraised that is not captured by the beta. Specific property risks include being a single stand-alone plant versus multiple plants in diverse industries, small size versus the competition, or illiquidity.

As pointed out above, the CAPM is one method used to calculate equity return. It is based on the concept that any stock's required rate of return is equal to the risk-free rate of return, plus its risk premium. The model is shown in table 5.5:

$$R_e = R_f + B(R_m - R_f) + R_u$$

Where:

R_e	=	Investor's required return on equity for the subject property.
R_f	=	The risk-free rate.
B	=	The market-value weighted average beta of the comparative property stocks.
$R_m - R_f$	=	The expected return of the market in excess of the risk-free rate, the market risk premium.
R_u	=	The additional risk premium.

Table 5.5. The Capital Asset Pricing Model

The basic model is shown in table 5.5 above. The main advantage of using the CAPM method is that it is largely derived from the marketplace. It is, therefore, less subjective than the build-up methods. The CAPM method's disadvantage is that it is based on certain assumptions that may be considered unrealistic. For example, the CAPM assumes there are no transaction costs, no taxes, and near-perfect market liquidity. Even with its shortcomings, however, the CAPM method remains a frequently employed method for determining the investor's required return on an equity investment.

Cost of Debt

Debt is incurred when the business borrows money from a capital provider. There is usually a specified interest rate and a specified term for repayment. The cost of the debt to the business is a function of the perceived risk of the business's ability to repay the debt. Like the cost of equity, the cost of debt has three components: a risk-free rate, an inflation factor (to account for the loss in value of the dollar due to inflation over the term of the debt), and a premium for risk.

In determining the cost of debt, the appraiser must consider both the market cost (a function of price and interest rate) and the marginal cost (the cost on the next dollar borrowed). It is also important that the "after-tax" cost of debt is used, since the interest payments on debt are usually tax deductible.

If an industrial plant is the subject property, the appraiser should note that, as a "small business," it will have greater difficulty obtaining debt financing than large companies, especially large companies that have equity traded on a stock exchange. One way to develop the market cost of long-term debt for such a "small business" property is to start by considering the yields on traded, rated debt such as that assigned a Standard & Poor's BBB bond rating, that in theory may reflect the risk of an investment in the subject property. However, depending on the size and other risk characteristics of the subject facility, the appraiser should remember that this rate may not be available to a potential investor in the subject property (i.e., the cost of debt could be higher).

In summary, the value of any asset is represented by the *present value* of its expected future benefits. The foundation of the income approach to value is to determine the present value of the expected future benefit from an asset, group of assets, or industrial property. To determine the present value, the MTS appraiser must understand the concepts of *time value of money* (present and future value), *discounting* and *compounding,* and *annuities.* The appraiser must also understand the *rate of return,* and that every investment, whether in land, buildings, or machinery and equipment, is made with the expectation of obtaining a return both of and on the investment. In addition, the appraiser needs to understand the *discount rate* and the various methods of determining it. These basic concepts form the income approach model that will now be discussed.

The Basic Steps

There are eight basic steps in the income approach:

1. Identify the expected future benefits.
2. Determine the magnitude and timing of the future benefits.
3. Determine the magnitude and timing of the expenses associated with achieving the forecasted income.
4. Subtract the annual operating expenses from the annual income or revenue.
5. Determine the discount rate (or capitalization rate if a multiple-year model is not being used).
6. Estimate the salvage or terminal value of the asset.
7. Determine the present value of the annual net cash flow.
8. Develop the appropriate value of the asset.

The first step in the income approach, as it pertains to machinery and equipment, is to *estimate the expected future benefits*. These are usually in the form of the specific income generated by the asset. However, in many cases the only income stream you can identify relates to the entire business, not the specific asset. If the asset or group of assets does not generate any income (e.g., the asset(s) is not in use, is not capable of being used) or if an income stream related to the assets cannot be identified, then it is important to note that it may not be possible to appraise the machinery and equipment using the income approach.

If an income stream can be associated with the asset or group of assets, then the next step is to *determine the magnitude and the timing of the future benefits*. This usually involves forecasting the total revenues generated by the asset, annually, over its remaining useful life. Historical income generated by the asset or forecasts prepared by the asset's owner can be used in this process. Integral to this step is determining the estimated remaining useful life of the asset. This is critical in the process and can have a significant impact on the value conclusion. In addition, when reviewing the historical operations of the subject, the historical utilization or capacity factor (actual production divided by maximum design production) should be studied, since the appraiser may want to forecast a normalized or stabilized level of performance that eliminates peaks or valleys in the revenue forecast.

When the income forecast is complete, the next step is to *determine the magnitude and timing of the expenses associated with achieving the forecasted income*. These expenses can be variable or fixed, direct or indirect, discrete or continuous, cash or noncash. Determining these expense categories and then the relevant costs frequently involves significant effort and can be a subjective effort. Interest and debt repayments are not included as expense items.

Next is to *subtract the annual operating expenses from the annual income or revenue*. The result is the annual net income that the asset will generate over its estimated remaining life. It is typical to determine a *net cash flow* after taxes by subtracting income tax from pre-tax operating income to obtain net income, adding back depreciation and other noncash expenses, and deducting capital expenditures and incremental changes in working capital (*working capital* is current assets minus current liabilities). If interest expense has been previously deducted (it should not have been, as pointed out above), then it must be added back, net of the tax deduction resulting from interest as a tax-deductible expense.

When the net cash flow forecast is complete, the *discount rate* (*or capitalization rate* if a multiple-year model is not being used) is estimated. The key to this step is analyzing the relative risk associated with the both the asset and the forecast and ensuring the risk is reflected in the rate that is selected. This is one of the main areas where income approach valuations are challenged. The appraiser should select the rate carefully and be able to justify it.

The next step in the basic model is to *estimate the salvage, residual, or terminal value of the asset*. Assuming the asset has a limited life, the method used to arrive at salvage value will determine whether you include salvage value in the next step, which is to *determine the present value of the annual net cash flows*. Depending on how you estimated the salvage value, either it may be in your final cash flow—and therefore it needs to be included in your present value calculations—or you may have estimated its present value in your analysis and therefore it is added to the present value of the annual cash flows. If the asset is actually a business such as an industrial plant whose life may be very long, then the terminal value is the value in the future capitalized into perpetuity. This terminal value is then discounted back to the appraisal date and added to the present value of the annual cash flow.

The final step—*to develop the appropriate value of the asset*—is dependent on the specifics of the valuation. For example, if the valuation is for a complex facility such as an industrial plant, the cash flow forecasts include both the returns "of" and "on" all the assets of the plant. This would include land, buildings, machinery and equipment, working capital, and any intangible assets. To estimate the value of the machinery and equipment, it is necessary to subtract the values for these other assets. On the other hand, if the valuation is for a property that is typically leased, and an asset-specific cash flow is developed, then once the discounted cash flow is determined and salvage value is added, the valuation process is complete.

When these basic steps are successfully performed, the result is an indication of value using the income approach. The actual application of this approach can be more complicated than indicated in this basic model. Keep in mind, however, that no matter how complex the situation, the basic concept remains the same.

The following section presents a relatively simplified example of the income approach as it applies to the valuation of an industrial facility. The reader will no doubt notice a distinct change from the discussion up to this point. So far, discussion has been centered on basic financial theory, introducing important terms and providing a "conceptual" income approach model. In the following section, there is an introduction to more specific terminology, demonstration of the theory with an actual cash flow model, and derivations of specific inputs into that model.

It is virtually impossible to present a universal model to appraise machinery and technical specialty assets using an income approach, because the MTS category is broad and every asset or facility is unique. This basic model will require adjustment for each appraisal situation in order to be properly applied. By understanding the theory and following the basic application model, the appraiser can make the proper modifications.

Application of the Income Approach

As we have seen, the income approach determines the present worth of future benefits (income) of ownership. **It is not usually applied to individual items of machinery and equipment unless they are leased.** However, it is frequently used to value a group of assets or individual machine units that are utilized together to produce a marketable product and that, in aggregate, generate an income stream. This collection of assets, along with any intangible and working capital assets, is frequently referred to as a business enterprise or going concern. A *business enterprise* is defined as all tangible assets (property, plant, equipment, and working capital) and intangible assets of a continuing business. In terms of value of the total invested capital, it is the combination of the value of stockholders' equity and the enterprise's long-term debt.

Two techniques are typically used to value machinery and equipment by the income approach, (1) the *direct capitalization approach* and (2) the *discounted cash flow (DCF) approach*. The direct capitalization approach measures value by dividing a projected income stream, in constant dollars, by a capitalization rate. The DCF approach is a method of analysis in which the quantity, variability, timing, duration of periodic income, and residual value are projected and discounted to a present value using a discount rate.

These approaches establish the value of an asset or group of assets assuming a going concern or business. A going concern or business implies that the operating assets being valued will remain in place and in use as part of a continuing and operating unit or business at their highest and best use.

To apply the income approach, the forecasted income stream must be positive. A positive income stream indicates that the business enterprise is a going concern, with future benefits of ownership. If the forecasted income stream is zero or negative, the assets must be valued under a premise of removal. This implies that the business is either losing money or, at best, breaking even. To maximize the assets' value, in theory, they should be deployed elsewhere. Examples of these two techniques are shown below.

Direct Capitalization Approach

The direct capitalization approach (also called the single-period model) simply capitalizes a projected net income or cash flow (expected future benefits) into perpetuity. It assumes that there will be no variation in the capitalization rate and no termination of the income stream (income into perpetuity). Two steps are required: (1) the income must be projected, and (2) the capitalization rate must be derived.

To project the income, the market for similar assets must be investigated and analyzed to locate comparable leased assets that can be adjusted to reflect a projected income for the subject asset. Appropriate adjustments may be made for such things as time, location, size, and age.

To derive the capitalization rate, sales of similar leased assets must be investigated and analyzed. The capitalization rate can be calculated by dividing the income of the sold asset by the sales price to determine the rate. This rate is frequently referred to as the overall rate because it captures all variables and risks associated with a particular income stream. The formula is stated as follows:

$$\frac{\text{Income}}{\text{Price}} = \text{Capitalization Rate}$$

After reviewing the comparable rates obtained from the market investigation, and adjusting them to the subject asset(s), a capitalization rate is developed. This rate can then be applied to the income projection for the subject asset(s) as follows:

$$\frac{\text{Subject Income Projection}}{\text{Capitalization Rate}} = \text{Income Indicator of Value}$$

Example

The subject machine is frequently leased in the marketplace; hence, an income approach is believed to be appropriate. The appraiser ascertained, from calls made to bankers, used equipment dealers, manufacturers, and other appraisers, that the following annual leases had been signed recently:

Last month	$1,000
Last week	$1,000
Last year	$970

The first two leases are at the same terms as the subject's, while the third lease is at a lower rate. Assuming all the comparable machines are identical to the subject, it appears that the third comparable lease needs to be adjusted upward (time adjustment). If price increases were about 3 percent, the adjusted lease rate is \$999 (\$970 × 1.03). Hence, an income of \$1,000 per year is a reasonable projection for the subject machine unit. Note that removing 3 percent for inflation from the discount rate implies that the annual income will increase at the rate of 3 percent (i.e., rental rate changes move at a constant 3 percent) and this must be verified during the investigation.

A similar investigation was performed to derive the capitalization rate. Two sales of similar machines were located, as illustrated in table 5.6:

Sale	Date	Sale Price	Annual Income	Derived Cap Rate*
1	Last week	\$5,300	\$800	15.1%
2	Last week	\$7,300	\$1,200	16.4%

* Derived Cap Rate = Annual Income/Sales Price

Table 5.6. Capitalization Rates Derived from Sales.

The two sales reviewed in this analysis reflect a range of 15.1 to 16.4 percent and a mean of 15.8 percent. Hence, 15.5 percent, slightly lower than the mean, was concluded to be reasonable. (Note: This was a subjective conclusion based on the appraiser's experience and judgment.)

Using the above data to develop an income indicator of value for the subject machine unit results in the following:

$$\frac{\text{Subject Income Projection}}{\text{Capitalization Rate}} = \frac{\$1{,}000}{15.5\%} = \$6{,}452$$

Hence, the income approach results in a rounded value indicator of \$6,500.

Capitalization rates may also be derived using the build-up method, the WACC, or the band of investment method by employing the Gordon-Shapiro model: discount rate (WACC) less projected growth rate.

Discounted Cash Flow Method

The DCF method (also called the multiple-period model) is most frequently developed on a debt-free, net cash flow basis. This recognizes the future cash outflows that are necessary to achieve the projected cash flows. The DCF technique results in the value of the business enterprise associated with a going concern.

This technique measures the direct economic benefits derived from ownership, in the form of future cash inflows and outflows attributed to the property, stated at their present value. Cash inflows are derived from income plus noncash expenses (depreciation expense). Cash outflows arise from future operating and general/administrative expenses, future capital expenditures, and any required influxes of working capital necessary to support growth and sales revenue.

Schedule of a Typical DCF Model

Revenues

Gross sales revenue from operations. It must not include interest income or other incomes from nonoperating assets. Revenues may be projected based on past historical results, in-house projections, published industry data, or a combination of these.

Cost of Goods Sold

All manufacturing and product-related costs and expenses, including raw materials, energy, manufacturing labor, maintenance labor, and similar variable expenses (this list is intended to be illustrative rather than exhaustive). Projections may be based on past historical results, in-house forecasts, published industry data, or a combination of these.

Gross Margin

Calculation: Revenues less cost of goods sold.

Operating Expenses

As used in this chapter, the term "operating expenses" includes all other expenses associated with achieving the projected revenues that are not included under Cost of Goods Sold. Included as operating expenses are such time-related or fixed expenses as depreciation, sales and marketing, general and administrative, corporate overhead, rent and lease, property taxes, and insurance (this list is intended to be illustrative rather than exhaustive). Interest on long-term debt is not an expense item and should not be deducted in arriving at net cash flow; otherwise, only equity will be valued, and not the total invested capital. Projections may be based on a review of past operating results, forecasts, published industry data, or a combination of these.

Operating Income

Calculation: Gross margin *less* total operating expenses.

Earnings Before Interest, Taxes, Depreciation & Amortization (EBITDA)

Calculation: Operating income *plus* depreciation expense. This is a useful level of income to compare against, because EBITDA excludes the effects of interest expense and debt repayment, depreciation, and the tax structure of the subject facility and its owner company. Sometimes referred to as the "net margin," this level of income is frequently compared against similar facilities to check or verify the reasonableness of historical operating results or projections.

Income Taxes

Composite federal and state income taxes.[3]

Net Income

Calculation: Operating income *less* income taxes (assuming "operating expenses" are defined as above); net income is sometimes expressed as pre-tax income less income taxes.

Depreciation

Adding back depreciation to net income equals what is sometimes called "gross cash flow."

Future Capital Expenditures

It is essential that future capital expenditures be sufficient to support the growth in revenue, earnings, and cash flow that are being projected. Estimating the capital expenditures required to support a given set of growth assumptions is an advanced topic that is discussed in the references given in the footnote.[4] If real or inflationary growth is projected during the specific forecast or terminal period, the projected growth will have to be supported by an excess of capital expenditures over depreciation. If zero growth is assumed, the appropriate assumption is to equate capital expenditures with depreciation.

Working Capital Changes

This is the amount the business invested in operating working capital. It is generally calculated based on a percentage of revenue change. If revenues are increasing, working capital additions will be positive, and if decreasing, negative, under the theory that if a business is growing it requires additional working capital to help sustain that growth. Conversely, if the business is shrinking, it will require less and less working capital.

Debt-Free Net Cash Flow

Calculation: Net income *plus* depreciation *less* capital expenditures *less* (or *plus*) net working capital changes. This is sometimes referred to as "free cash flow." Net cash flow is the total after-tax cash flow generated by the going concern and available to the providers of the facility's investment capital: stockholders (equity) and creditors (debt).

The above schedule represents the basic model that is used to restate the facility's actual historical operating statements and to forecast the future in a DCF analysis. The number of years included in what is called the "specific forecast" period is based on several factors, such as the economics of the subject industry and the economics and the physical attributes of the subject property. If the subject assets are physically very old and obsolete, the remaining life of the property may be very short. While it is possible to spend large amounts of money to rebuild a facility, it may not be economically prudent. Hence, the forecast period in such a case could be very short and the final year would represent facility shutdown and the liquidation of its assets (salvage or terminal value).

In cases where the property has a long remaining life, the DCF could theoretically be forecast at 100 years, but after 30 years or so, the present value of the net cash flow is so small that the exercise is moot. Generally, after the changes in net cash flow begin to stabilize (for example, after 5, 10, or 15 years) the net cash flow stream is capitalized into perpetuity.[5]

Discount Rate Derivation

The estimation of an appropriate discount rate or WACC is fundamental to any computation of present value. (See basic step 5.) The following steps are employed to develop the discount rate:

- capital structure analysis;
- cost of debt analysis;
- cost of equity analysis; and
- calculation of the discount rate or WACC.

As part of the analysis, companies in the subject industry are studied (in business valuation these are called "guideline companies"). Although these companies may not be entirely similar to the subject because of size and other operating characteristics, they are generally affected by the same economic factors and are viewed as reasonably comparable for the purpose of estimating the cost of capital.

Capital Structure Analysis

The comparables of a sample analysis are listed below in table 5.7, with long-term debt and equity expressed as a percentage of total capital structure, and working capital as a percentage of revenues

Company	WC as a % of Revenue	Long-Term Debt	Equity
Company A	10%	45%	55%
Company B	12%	40%	60%
Company C	8%	35%	65%
Company D	9%	40%	60%
Company E	10%	45%	55%
Company F	11%	40%	60%

Table 5.7. Capital Structure of Comparable Companies

This type of data is available from various sources such as the *Value Line Investment Survey*, Standard & Poor's, annual stockholder reports, SEC filings, and other published sources. The typical capital structure for the subject industry was developed from the average capital structure of the guideline companies. Long-term debt and equity must be at market rates, not as reported by accountants.

The appropriate guideline company capital structure was determined to be 60 percent equity and 40 percent debt. This may or may not be similar to the actual capital structure of the subject property. If the subject property is a small division of a large corporation, it may not have a capital structure and may contribute only a small income stream to the company. In this case, the large corporation may have a capital structure that is not reflective of a company in the subject's industry.

Cost of Debt Analysis

Stockholders and debt holders require compensation for investing in a property comparable to the subject versus one of similar risk. This compensation is often called the opportunity cost. The cost of equity is equivalent to the return required by an equity investor from an investment comparable to the subject property. The cost of debt is the required cost of interest on debt for a loan on a property comparable to the subject.

After determining the capital structure, the rates of return required in the capital markets are reviewed. The market rates, reported by Standard & Poor's *Monthly Bond Guide,* Moody's *Bond Guide,* the *Wall Street Journal* and many others, are summarized below in Table 5.8, and represent the average rates as of the appraisal date.

	Yield (%)
Prime Rate	6.00
Short-Term Government Bills	5.13
Intermediate-Term Government Notes	5.63
Long-Term Government Bonds	6.74
Industrial AAA Bonds	7.29
Industrial BBB Bonds	8.40

Table 5.8 Rates of Return Required in Capital Markets

A small business such as the subject facility can be expected to have greater difficulty obtaining debt financing than larger companies, especially those with equity traded on a stock exchange. One way to develop the market cost of long-term debt for the subject is to start by considering the yields on traded, rated debt, such as debt with S&P's BBB bond rating, which had a 8.4 percent yield on the appraisal date. However, depending on the size and other risk characteristics of the subject facility, this rate may not be available to a potential investor in the subject. Although 8.4 percent has been used as the cost of debt in the following examples, it should be noted that using yields on traded, rated debt might understate the cost of debt for a small business such as the subject facility. Quantifying the additional risk is a complex issue (see the reference cited in the footnote).[6]

Cost of Equity Analysis

The CAPM method may be used to relate the return that equity investors require to the risk-free return offered by government securities. To account for relative risks for specific industries, the additional returns or risk premium required by the market in general can be adjusted by a beta factor (ß). The CAPM uses the following formula to arrive at an appropriate cost of equity:

$$R_e = R_f + \beta (R_m - R_f) + R_u$$

Where:

R_e = Investor's required return on equity for the subject property.

R_f = Risk-free rate.

β = Market value weighted average beta of the guideline company stocks.

$R_m - R_f$ = Expected historical return of the market in excess of the historical risk-free rate.

R_u = Additional risk factor.

The risk-free rate, as approximated from the yields on long-term U.S. Treasury securities, was estimated at 6 percent. This is based on the yields on intermediate (10-year) and long-term (20-

to 30-year) government securities (see table 5.8: 5.63% for intermediate government securities and 6.74% for long-term government securities).

The expected historical return of the market in excess of the historical risk-free rate has been estimated at 7.1 percent; this figure is based on a study of actual returns, according to *Stocks, Bonds, Bills, and Inflation: Yearbook* by Ibbotson Associates.

The beta component of the equation reflects the risk of the specific investment relative to the return of the overall stock market. The investigation included a review of betas for companies in the subject industry, as reported by the *Value Line Investment Survey*. The analysis indicated that the companies reviewed generally have betas between 0.80 and 1.15, with the average beta about 0.98. Hence, for this analysis, a beta of 1.0 was chosen.

The final component of the equation is the additional risk factor. Because the subject property is a single stand-alone facility, for purposes of this example we will assume that an additional risk factor of 4 percent is appropriate.

Employing the CAPM methodology, the cost of equity *(R_e)* was estimated as follows:

$$R_e = R_f + \beta(R_m - R_f) + R_u$$
$$R_e = 6.0 + 1.0\ (7.1) + 4.0$$
$$R_e = 17.1\%$$

The 4% additional risk factor added to the cost of equity calculated above is sometimes called the small-company risk premium, the specific-company adjustment, or the illiquidity premium. Unless the subject property is a large, liquid, publicly traded business, the cost of equity calculated by the basic CAPM model is only a starting point in estimating the cost of equity for an industrial facility. Note that in the example that follows, the subject property is assumed to be something other than a large, liquid, publicly traded company.

Calculation of Discount Rate or WACC

The discount rate or WACC is then calculated by weighting the cost of debt (tax-affected at 40 percent[7] to reflect the deductibility of interest expense) and the cost of equity by the industry capital structure indicated in an analysis of comparable companies. In the example, the typical market-based, industry capital structure was assumed to be composed of 40 percent debt and 60 percent equity. Therefore, a WACC of 12.3 percent was calculated, as shown in Table 5.9.

	1 Capital Structure	2 Pretax Cost	3 Tax Effect*	4 After-Tax Cost	5 (1 x 4)
Debt	40%	8.4%	60.0%	5.04%	2.02%
Equity	60%	17.1%	100.0%	17.10%	10.26%
WACC					12.28%

* The tax rate is assumed to be 40%

Table 5.9 Weighted Average Cost of Capital

This discount rate is intended for application to an after-tax, debt-free net cash flow. At this level of cash flow, interest earnings on other investments are not added to revenues, interest expense is not an expense item, and debt repayment is not a deduction.

Occasionally, certain states and financial institutions require net cash flow streams to be pre-tax, pre-depreciation, or pre-property tax. Therefore, the discount rate must be adjusted to reflect the nature of the income to be discounted. It is sometimes difficult to make the appropriate adjustments, but if made correctly the final result will not be affected.

The subject property's historical operating statements for the past five years are shown in table 5.10.

	5 Years Ago	4 Years Ago	3 Years Ago	2 Years Ago	1 Year Ago
Revenue	84,935	88,474	92,160	96,000	100,000
– Cost of Goods Sold	(53,118)	(54,760)	(56,454)	(58,200)	(60,000)
= Gross Margin	31,817	33,713	35,706	37,800	40,000
– Operating Expenses	(17,706)	(18,253)	(18,818)	(19,400)	(20,000)
– Depreciation	(3,453)	(3,559)	(3,670)	(3,783)	(3,900)
= Operating Income	10,659	11,900	13,218	14,617	16,100

Table 5.10 Historical Operating Statements

Income taxes are not shown as an expense item because they are based on a specific corporate structure and do not reflect the marginal statutory rate of the subject property. These taxes will be deducted in the DCF model that follows.

The simplified, comparative historical operating statements shown in table 5.10 are a starting point in identifying the expected future benefits, as outlined in basic step 1. The following section discusses basic steps 2 and 3, the determination of the magnitude and timing of the future benefits, and the costs of achieving the forecasted income stream.

Historical trends in the industry, future projections of industry economics, and the subject property's past operating results are used to project future operating results. The major assumptions, projections, estimates, and calculations used in the business enterprise valuation for the subject property are explained as follows:

- Revenues and costs of goods sold are projected to increase or decrease based on short- and long-term projections for the industry. Primary industry projections were based on data published in industry publications and consultants reports.
- Operating expenses are based on historical operating results and are estimated to increase based on historical growth. Corporate overhead may be restated to industry norms.
- Operating income is the gross margin net of cost of goods sold. Income taxes are calculated based on the composite state and federal rate of 40 percent. Net income is operating income less income taxes.
- Net cash flow is net income plus depreciation expense, minus future capital expenditures, minus (or plus) working capital changes (basic step 4). Working capital requirements can be based on a review of the working capital requirements of comparable companies or specific requirements of the subject facility. The amount of future capital expenditures was based on the long-term requirements of the facility. To complete basic step 4, subtract expenses from revenue (basic step 5 was previously discussed in the discount rate derivation section of this chapter).

The cash flows are discounted to present worth at an appropriate discount rate over the ten-year projection period (basic step 7). For the residual period beyond year ten, the projected cash flow is capitalized into perpetuity employing the Gordon-Shapiro model, the denominator in the fraction being the discount rate less the cash flow rate of growth. The resulting capitalized value is converted to present worth using the present value factor from year ten (basic step 6). Note that present value factors for the interim years are taken at the end of the year.[8]

The value of the subject property as a viable operating entity into the future is equal to the sum of the present values of the projections of cash flow plus capitalization of the projected residual cash flow, the latter of which is also discounted to present value (basic step 8).

Table 5.11 illustrates a simplified income approach for the subject property.

The cash flow for the last year of the specific forecast period (year ten in this example) is $10,707 and is capitalized into perpetuity using a capitalization rate derived from the discount rate using the Gordon-Shapiro model.

If zero growth into perpetuity is assumed, the capitalization rate for converting the $10,707 cash flow into a residual value is the same as the discount rate of 12.3 percent (12.3% – 0% growth = 12.3%). If the appraiser were assuming growth in cash flow of 2 percent into perpetuity, the numerator in the residual value fraction would be $10,707 × 1.02 and the denominator would be 10.3 percent (12.3% – 2.0% = 10.3%). In the example, the residual value is calculated by dividing $10,707 by 12.3 percent ($10,707 × 12.3% = $87,045). This figure represents the terminal or residual value of the facility at the end of year ten/beginning of year eleven, assuming the facility will be operated into perpetuity (basic step 6). This future value must then be present valued back to the appraisal date. This is done using Formula #2, which multiplies the present value factor of 0.3135 (the present value factor for the tenth year at 12.3 percent) times the $87,045 future value. The result is a present value of $27,286. When this terminal value is added to the sum of the present value of each year's (years one through ten) cash flow, the result is the business enterprise value, which in the example is $88,764 (basic step 7).

As noted previously, a business enterprise has three components: working capital, tangible assets, and intangible assets. In the financial analysis, it was concluded that the appropriate working capital on the valuation date (year zero) was 10 percent of revenue. This would indicate a working capital of $10,000 based on the revenue on the valuation date (year zero) of $100,000.

Deducting the working capital from the business enterprise value results in an indication of value of $78,764 for the subject's tangible and intangible assets, which has been rounded to $79,000. This figure includes the value of all tangible and intangible assets, including machinery and equipment, real property (build-

	Year	1	2	3	4	5	6	7	8	9	10	Residual
	Revenue	105,000	110,250	113,558	116,964	120,473	122,883	125,340	127,847	129,126	130,417	
–	Cost of Goods Sold	(61,800)	(63,654)	(65,564)	(67,531)	(69,556)	(71,643)	(73,792)	(76,006)	(78,286)	(80,635)	
=	Gross Margin	43,200	46,596	47,994	49,434	50,917	51,239	51,548	51,841	50,839	49,782	
–	Operating Expenses	(20,600)	(21,218)	(21,855)	(22,510)	(23,185)	(23,881)	(24,597)	(25,335)	(26,095)	(26,878)	
–	Depreciation	(3,900)	(4,095)	(4,218)	(4,344)	(4,475)	(4,564)	(4,655)	(4,749)	(4,796)	(4,844)	
=	Operating Income	18,700	21,283	21,921	22,579	23,257	22,794	22,295	21,757	19,948	18,059	
–	Income Tax	(7,480)	(8,513)	(8,769)	(9,032)	(9,303)	(9,118)	(8,918)	(8,703)	(7,979)	(7,224)	
=	Net Income (Debt Free)	11,220	12,770	13,153	13,547	13,954	13,677	13,377	13,054	11,969	10,836	
+	Depreciation	3,900	4,095	4,218	4,344	4,475	4,564	4,655	4,749	4,796	4,844	
–	Capital Expenditures	(5,243)	(5,505)	(5,670)	(5,840)	(6,015)	(6,136)	(6,258)	(6,384)	(6,447)	(4,844)	
–/+	Changes in Working Capital	(500)	(525)	(331)	(341)	(351)	(241)	(246)	(251)	(128)	(129)	
=	Debt Free Net Cash Flow	9,377	10,835	11,370	11,711	12,062	11,864	11,528	11,168	10,189	10,707	10,707
	Present Value Factor	0.8905	0.7929	0.7061	0.6288	0.5599	0.4986	0.4440	0.3953	0.3520	0.3135	12.3%
	Present Value	8,350	8,591	8,028	7,363	6,754	5,915	5,118	4,415	3,587	3,356	87,045
	Sum of PV of Cash Flows in Years 1–10		61,478									
+	Residual or Terminal Value		27,286									
=	Value of Business Enterprise		88,764									
–	Working Capital		(10,000)									
=	Value of Tangible and Intangible Assets		$78,764									
	Rounded		$79,000									

Table 5.11. Sample Income Approach

ings, other structures, site improvements, and land), personal property, and intangible assets.

If the objective of the appraisal requires the exclusion of intangible value, the appraiser may need to separately value the intangibles if they are deemed a material part of the total value derived above. The intangible component of value may or may not be significant to the accuracy of the appraisal, depending on the facility's level of profitability and other factors.[9] If the appraiser needs to bring the income approach all the way down to the value of the machinery and equipment, his or her task will be to determine the value of the real property and intangibles and deduct these from the value of the tangible and intangible assets.

Key Points

- A decision to purchase any property is an investment decision. Investment decisions are made in consideration of the expected return to be generated, the period of time over which the return will be earned, and the risk of not receiving the expected return.
- The income approach to value is a method to measure the present value of the property's expected future returns.
- The time value of money concept is that a dollar received today is worth more than a dollar to be received in the future, because the dollar received today could be invested and earn interest. Present value represents the value today of something that will be received in the future. For example, if you were going to receive one dollar five years from now, what would that future payment be worth if you received it today instead? The answer is it would be worth the amount of money that would need to be invested today in order to have one dollar five years from now, at a specified interest rate or rate of return. The process used to determine the present value is referred to as *discounting.*
- Investments are generally made with two basic goals in mind: achieving the return "of" the original investment—getting the original investment back—and a return "on" the investment—getting more than the original investment back. It is important in the investment decision process to make investments that will provide a rate of return that is commensurate with the risk

of the investment. It is this return, in the form of income or cash flow, that is generated by the asset, that is quantified and discounted to determine value using the income approach.

- The discount rate is the rate used in discounting the future benefits generated by an asset to determine present value. When someone buys a piece of machinery or equipment, he or she is making an investment. He or she expects the equipment to provide a return both "of" and "on" the investment, usually in the form of income or cash flow. In determining the value of a piece of equipment using the income approach, the present value of the expected future benefits from that equipment must be determined.
- The *weighted average cost of capital* is an appropriate *discount rate* when valuing the assets of a business by the income approach. A company's cost of capital is the return that it must provide to investors for additional capital, that is, short-term debt, long-term debt, and preferred and common stock (equity).
- The most common methods for estimating the investor's required return on equity are the capital asset pricing model (CAPM) method and the build-up method.
- Two techniques are typically used to value machinery and equipment by the income approach: the direct capitalization approach and the DCF approach. The direct capitalization approach measures value by dividing a projected income stream, in constant dollars, by a capitalization rate. The DCF approach is a method of analysis in which the quantity, variability, timing, duration of periodic income, and residual value are projected and discounted to a present value using a discount rate.
- The direct capitalization approach simply capitalizes a projected net income or cash flow into perpetuity. It assumes that there will be no variation in the capitalization rate and no termination of the income stream. It is a single-period model.
- The DCF method is most frequently developed on an invested capital, net cash flow basis. This eliminates the effects of how the actual business is currently being financed and taxed and also recognizes the future cash outflows that are necessary to achieve the projected cash flows. The DCF method results in

the value of the business enterprise associated with a going concern. It is a multiple-period model.

- The discount rate used in the DCF method may be estimated by determining the subject property's weighted average cost of capital (WACC). It is calculated by weighting the cost of debt and the cost of equity.
- This chapter has examined a complex topic. The reader is referred to business valuation literature, real estate valuation literature, and investment texts for further study.
- The income approach is a practical, useful tool that may be applicable to certain machinery and equipment appraisal projects.

Additional Reading

Brealey, Richard A., and Stewart C. Myers. *Principles of Corporate Finance.* 5th ed. New York: McGraw Hill, 1998.

Copeland, Tom; Tim Koller; and Jack Murrin. *Valuation: Measuring and Managing the Value of Companies.* 2nd ed. New York: Wiley & Sons, 1995.

Cornell, Bradford. *Corporate Valuation*. Homewood, Ill.: Business One Irwin, 1993.

Pratt, Shannon P. *Cost of Capital: Estimation and Applications.* New York: John Wiley & Sons, Inc., 1998.

Pratt, Shannon P.; Robert F. Reilly; and Robert P. Schweihs. *Valuing a Business.* 3rd ed. Chicago: Irwin Professional Publishing, 1996.

Notes

[1] Shannon P. Pratt, *Cost of Capital: Estimation and Applications* (New York: John Wiley & Sons, Inc., 1998), p. 21.

[2] Pratt, *Cost of Capital: Estimation and Applications.*

[3]Composite Income Tax Rate = Federal Rate + (State Rate [1 – Federal Rate]).

[4] See the discussion of this issue at pages 156-160 of *Corporate Valuation* by Prof. Brad Cornell; also see pages 325–328 of *Analysis for Financial Management* by Prof. Robert C. Higgins. 5th ed. Boston, Mass.: Irwin/McGraw-Hill. 1998.

[5] Unlike corporations, tangible assets have finite useful lives; thus capitalizing the terminal cash flow into perpetuity may not be the appropriate assumption, especially if the facts indicate that the subject tangible asset has a relatively short remaining useful life. In this case, the appraiser may decide to forecast each year's cash flow during the asset's estimated remaining life and then forecast the salvage value, if any, that would be realized at the end of the asset's life.

For additional guidance, see Hal B. Heaton, "Valuing Small Businesses: The Cost of Capital," in *The Appraisal Journal* (January 1998), p. 12. In addition to the discussion in the above article, it may be possible for the appraiser to identify debt rates actually available to small companies similar in size and risk to the subject facility.

[6,7] Composite federal and state income tax rate.

[8] While end-of-year present value factors are used in this example, midyear present value factors, or factors representing a continuous flow of funds, can also be used.

[9] For additional guidance on the valuation of intangibles, the appraiser may wish to review Gordon V. Smith and Russell L. Parr, *Valuation of Intellectual Property and Intangible Assets*, 2nd ed. (New York: Wiley & Sons, 1994); and Robert F. Reilly and Robert P. Schweihs, *Valuing Intangible Assets,* New York: McGraw-Hill, 1999).

6

Report Writing

Objectives:

1. Describe the purpose and use of appraisal reports.
2. State pertinent rules and standards applicable to MTS appraisal reports.
3. Relate content considerations for appraisal reports.
4. Define the parts of appraisal reports.
5. Provide report organization and writing tips.

The appraisal report—like painting, sculpture, or architecture—is an art form. It should be pleasing in appearance and precise in content. It should take the reader by the hand and lead him along the line of thought adopted by the writer toward the goal of a convincing conclusion. Along the way, the reader should find satisfaction, education, and evidence of skillful use of knowledge. In addition, the report should be professional in design, logical in reasoning, and completely convincing in text.[1]

This chapter is designed to assist the appraiser in successfully preparing a coherent and comprehensive appraisal by (1) providing the minimum requirements and the 5 general reporting requirements outlined by USPAP and (2) detailing the when and how of reporting, including organizing the work and writing the report. This chapter will also review effective writing techniques and the various reporting styles.

Purpose and Use of Appraisal Reports

Documenting the appraisal work is the culmination of the appraisal investigation. The purpose of the appraisal report is to identify the property, organize the data, describe the appraisal process, and explain what the appraiser found and how conclusions were obtained. Like a story, the report needs a beginning, middle, and end, but the report should also conform to a certain set of norms and reporting conventions. The report writer should visualize the report as a structure. The report guides both the writer and the reader to a clear statement of the work; it serves as a map for the reader and tells the reader what to expect.[2]

Writing the Report

Getting started is an anxious moment for veteran and neophyte appraisers alike. However, documenting the work is the culmination of the appraisal investigation.[3] The report identifies the property and organizes the data. It explains what the appraisers found, and how they came to their conclusions. Report writing also follows a certain set of norms and reporting conventions.

When telling their story, appraisers often burden the reader with too much fluff and not enough substance. Visualize the report as a well-built structure. The foundation should be solid, the walls should be strong, and the roof should be sound.

Why Write Reports?

Reports are written for the appraiser as well as for the client. The report guides the writer and the reader to a clear statement of the work; it is also a kind of map for the reader and tells the reader what to expect. Appraisals are used to help make decisions for such things as financing, insurance, property tax purposes, allocation of purchase price, and litigation.[4]

Rules and Standards

USPAP contains the minimum standards for the development and reporting of an appraisal. For personal property such as machinery and equipment, professional appraisers are generally subject to all of USPAP Standards, but are more specifically governed by Standard 7 (personal property appraisal) and Standard 8 (personal property appraisal reporting).

This chapter is dedicated to report writing and, therefore, will discuss only Standard Rule 8. The minimum requirements for professional appraisal reporting under Standards Rule 8 are discussed below.[5] Departure from binding requirements is not permitted; each binding requirement is noted below.

> **Standards Rule 8-1**[6]
>
> Each written or oral personal property appraisal report must:
>
> (a) clearly and accurately set forth the appraisal in a manner that will not be misleading; **[Binding requirement.]**
>
> (b) contain sufficient information to enable the intended users of the appraisal to understand the report properly. **[Binding requirement.]**
>
> (c) clearly and accurately disclose any extraordinary assumption, hypothetical condition, or limiting condition that directly affects the appraisal and indicate its impact on value. **[Binding requirement.]**
>
> **Standards Rule 8-2**
>
> According to USPAP, "each written personal property appraisal report must be prepared under one of the following three options and prominently state which option is used: Self-Contained Appraisal Report, Summary Appraisal Report, or Restricted Use Appraisal Report."[7] For purposes of our discussion, we will look at Standard Rule 8-2(a), which states that

The content of a Self-Contained Appraisal Report must be consistent with the intended use of the appraisal and, at a minimum:

(i) state the identity of the client and any intended users, by name or type;

(ii) state the intended use of the appraisal;

(iii) describe information sufficient to identify the property involved in the appraisal, including the physical and economic property characteristics relevant to the assignment;

(iv) state the property interest appraised;

(v) state the purpose of the appraisal (the type and definition of value) and its source;

(vi) state the effective date of the appraisal and the date of the report;

(vii) describe sufficient information to disclose to the client and any intended users of the appraisal the scope of work used to develop the appraisal;

(viii) state all assumptions, hypothetical conditions, and limiting conditions that affected the analyses, opinions, and conclusions;

(ix) describe the information analyzed, the appraisal procedures followed, and the reasoning that supports the analyses, opinions, and conclusions;

(x) state, as appropriate to the class of personal property involved, the use of the property existing as of the date of value, and the use of the property reflected in the appraisal; and, when the purpose of the assignment is any market value, describe the support and rationale for the appraiser's opinion of the highest and best use of the property;

(xi) state and explain any permitted departures from specific requirements of STANDARD 7, and the reason for excluding any of the usual valuation approaches;

(xii) include a signed certification in accordance with Standards Rule 8-3.[8]

USPAP Standard 8-2(b) and 8-2(c) carry much the same requirements for the Summary Appraisal Report and the Restricted Appraisal Report as that described for the Self-Contained Appraisal Report. Advisory Opinion 11 (AO-11) explains the different report requirements as follows:

> The essential difference among the three options is in the use and application of the terms *describe, summarize,* and *state. Describe* is used to connote a comprehensive level of detail in the presentation of information. *Summarize* is used to connote a more concise presentation of information. *State* is used to connote the minimal presentation of information.[9]

As indicated, the appraiser should decide the proper type of appraisal report consistent with the intended use of the appraisal.

Standards Rule 8-3

Each written personal property appraisal must contain a signed certification that is similar in content to the following form:

I certify that, to the best of my knowledge and belief:

— the statements of fact contained in this report are true and correct.

— the reported analyses, opinions, and conclusions are limited only by the reported assumptions and limiting conditions, and are my personal, impartial, and unbiased professional analyses, opinions, and conclusions. I have no (or the specified) present or prospective interest in the property that is the subject of this report, and no (or the specified) personal interest with respect to the parties involved.

— I have no bias with respect to the property that is the subject of this report or to the parties involved with this assignment.

- — my engagement in this assignment was not contingent upon developing or reporting predetermined results.
- — my compensation for completing this assignment is not contingent upon the development or reporting of a predetermined value or direction in value that favors the cause of the client, the amount of the value opinion, the attainment of a stipulated result, or the occurrence of a subsequent event directly related to the intended use of this appraisal.
- — my analyses, opinions, and conclusions were developed, and this report has been prepared, in conformity with the Uniform Standards of Professional Appraisal Practice.
- — I have (or have not) made a personal inspection of the property that is the subject of this report. (If more than one person signs the report, this certification must clearly specify which individuals did and which individuals did not make a personal inspection of the appraised property.)
- — no one provided significant professional assistance to the person signing this report. (If there are exceptions, the name of each individual providing significant professional assistance must be stated.)[10]

In addition to USPAP requirements, the ASA has requirements pertaining to what an appraisal report must contain. In "Appraisal Reports," Section 8, of *The Principles of Appraisal Practice and Code of Ethics,*[11] the ASA describes what must be included in appraisal reports. These requirements are generally the same as those in USPAP, with the exception of Sections 8.7 and 8.8. These are as follows:

> **Section 8.7—Mandatory Reaccreditation Statement**
>
> All Senior Member appraisers should state in each report: "The American Society of Appraisers has a mandatory reaccreditation program for all of its Senior members. 'I am' or 'I am not' in compliance with that program."
>
> **Section 8.8—Signatures to Appraisal Reports and Inclusion of Dissenting Opinions**

> It is required that the party who makes the appraisal or who has the appraisal made under their supervision sign the appraisal report. (See Sec. 7.4.)
>
> It is required that all collaborating appraisers, issuing a joint report, who agree with the findings, sign the report; and that any collaborating appraiser who disagrees with any or all of the findings of the others, prepare, sign, and include in the appraisal report his dissenting opinion. (See Sec. 7.4.)

Appraisers should be aware that any report, oral or written, provided under USPAP, must comply with all of the above which are binding requirements, and that departure from them is not permitted.

The requirements are ethical and commonsensical. The Appraisal Foundation, through review and recommendation of certain appraisal policies, understands that all directives must be realistic and flexible. There are certain "departure" provisions that allow appraisers some latitude in collecting and reporting data. However, there are also "binding" provisions in the Standards that do not allow for departure. Because USPAP is an evolving set of directives, it is important that the appraiser is familiar with the current version of USPAP.

These requirements help appraisers organize data to clarify valuation problems and support their conclusions. Although tedious at times, this is not a difficult task. If the writer organizes the material with the final report in mind, there will be little trouble in helping the client understand the appraisal.

Content Considerations

Although not necessarily initially committed to paper, report writing begins with the start of the appraisal, when the appraiser determines the level of information needed and how to list and describe the property and assets. The report should be provided to the client within a contractually specified time period, and should always be completed immediately following the appraisal.

An appraisal report can have several audiences. The initial audience is usually the individuals who will use the information to facilitate their business. The primary client will normally have some expertise in the property appraised. However, there are times when the client, or a third party, has no understanding of the property appraised. Therefore, appraisal reports should be written to provide a basic understanding, regardless of client knowledge level. A well-written appraisal report is one that enables readers to fully understand what is being conveyed even if they have no prior knowledge of the property or the appraisal process.

About the Report

All appraisals, written or oral, must have all supporting documentation maintained in the appraiser's file. Therefore, a properly prepared appraisal report will be supported by the backup data. The record-keeping provision of USPAP states that "An appraiser must prepare a workfile for each assignment" and must "retain the workfile for a period of at least five (5) years after preparation or at least two (2) years after final disposition of any judicial proceeding in which testimony was given, whichever period expires last."[12] The backup data may include contacts, telephone logs, letters, and relevant data from any source that assisted with the appraisal.

Clients and potential clients often ask appraisers for "quick-and-dirty" estimates of value, or for verbal reports. The quickest and most libelous road to trouble is to provide an "unconsidered opinion." Consider again USPAP Standard Rule 8-1, which states that:

> Each *written or oral* personal property appraisal report must
>
> (a) clearly and accurately set forth the appraisal in a manner that will not be misleading;
>
> (b) contain sufficient information to enable the intended users of the appraisal to understand the report properly;
>
> (c) clearly and accurately disclose any extraordinary assumption, hypothetical condition, or limiting condition that directly affects the appraisal and indicate its impact on value.[13]

When to Write the Report

Logically, at the beginning of the appraisal, the appraiser will have already decided how to list the property, how to describe the asset, and how much information to include. Appraisers will have organized their thoughts and supporting backup material relative to the market and other contributing factors. As a practical matter, the report should be written in a reasonable time, so as to not allow the information to become outdated. The appraiser should write the report shortly after completing the valuation process, because clients will often need the report as soon as possible. Furthermore, appraisers usually have a contractual obligation to perform the appraisal in a certain period of time.[14]

To Whom Is the Report Written?

The writer should first determine the report's audience. The initial audiences are usually decision makers, the individuals who will use the information to facilitate their business. Appraisers may assume that their clients have some expertise in the property appraised; but there may be times when the client, or intended third party, is not familiar with the property. Therefore, the appraisal report should be written so that any privileged party can understand it.[15]

What Does the Reader Already Know About the Property?

The primary client may have total familiarity with the property. However, the report can include details that someone not familiar with the property will need to know. For example, an intended third-party reader may not know the relationship of certain machines to the overall manufacturing process. If the valuation is for *fair market value in continued use,* the report may need to explain such things as plant capacity, technological changes in relationship to overall age of the process, and so on.[16]

What Does the Reader Want to Know About the Property?

In addition to all of the required information, the reader will want to know all of the strengths and weaknesses of the property as well as the conspicuous facts relevant to value.[17]

What Does the Reader Need to Know About the Property?

Anything that will affect the value of the property, if known by the appraiser, should be considered in arriving at a value conclusion. Depending on its format, the appraiser may want to discuss in the report some or all of the factors affecting value. For example, the appraiser will consider, and may want to discuss in the report, that a machine is so outdated that its only viable use is at its present location. Similarly, the appraiser will consider, and may want to discuss in the report, that certain machinery is custom made for a particular facility and may have no potential use elsewhere.[18]

What Point of View Does the Reader Hold?

This may or may not be a consideration, as the report has to accommodate many users. Intuitively, the appraiser will grasp the importance that the client attaches to the property appraised. Although reports are straightforward, it may make a difference if the appraiser understands that the client is the business founder who knows the business intimately, or is an outside investor looking strictly for facts.[19]

What Are the Reader's Reasons for Reading the Document?

Users of appraisals need to be informed and to be able to use the report to make a decision. Some users of appraisal reports will read the summary and spend little time on the remainder. However, it is possible that the report will be thoroughly scrutinized by the primary client or third party, or possibly reviewed by another appraiser.

What Should the Reader Know or Be Able to Do After Reading the Document?

The reader should be able to comprehend the information and have a useful business tool. The report should be complete, objective, and factual.[20]

Other Factors/Standards Affecting the Report

The value reported should coincide with the definition of value and purpose of the appraisal and should be supported with proper documentation. USPAP Standard Rule 7-3 states the additional factors that need to be considered during the appraisal process.

> **Standards Rule 7-3**
>
> In developing a personal property appraisal, an appraiser must collect, verify, analyze, and reconcile all information pertinent to the appraisal problem, given the scope of work identified in accordance with Standards Rule 7-2(f).
>
> (a) Where applicable, identify the effect of highest and best use by measuring and analyzing the current use and alternative uses to encompass what is profitable, possible, legal, and physically possible, as relevant to the purpose and intended use of the appraisal;
>
> (b) Personal property has several measurable marketplaces, therefore the appraiser must define and analyze the appropriate market consistent with the purpose of the appraisal;
>
> (c) Analyze the relevant economic conditions at the time of the valuation, including market acceptability of the property as well as supply, demand, scarcity, or rarity.[21]

Standards Rule 7-4 requires a personal property appraiser to collect, verify, and analyze all information applicable to the appraisal problem, and if applicable, the appraiser must apply the cost, sales comparison, and income approach to value.

What Is the Primary Purpose?

The report's primary purpose is to document the factual, objective, and succinct information—the what, when, why, where, and how—of the appraisal assignment.

Organizing and Writing the Report

Report writing is a process of bringing the information, both written and oral, to the client in a usable form. William C. Himstreet, in his excellent monograph, *Writing Appraisal Reports,* compares the appraisal report process to an information funnel. He begins at the top of the top of the funnel with the research that includes the appraiser's data gathered into usable information segments.

Writing the Final Report

Specifically, the job is now to

- condense the material from original notes, cost surveys, and observations, and remove superfluous material;
- combine the written information including charts and surveys, if they are used;
- assimilate the usable information into a form that serves the reader's needs;
- write the report, making certain that it is not fragmented or disjointed; and
- ensure the report conforms to USPAP.

Appraisers can follow the requirements of USPAP while developing their own reporting formats. Some of the following guidelines are not necessarily required by USPAP, but may help the appraiser prepare a more complete appraisal report.

General Guidelines to Organizing and Writing the Report

Title Page

Although not a USPAP requirement, most good appraisal reports include a title page that lists the addressee, the name of the property appraised, and the client(s) for whom the appraisal is prepared. The title page can also include the effective date of the appraisal, the valuation concept, and the name of the appraiser or appraisal firm.

Cover Letter

Although USPAP does not require a cover letter, convention does. The cover letter, usually not more than one page, is a transmittal that conveys the appraisal conclusions to the client. The cover letter can be separate or bound within the report. This is a personal preference. An advantage of the bound letter is that it is less likely to be lost or become detached from the report; if that were to happen, the reader might be misled. It is important to reference the body of the report in the cover letter.

Index or Table of Contents

Just as it suggests, this is a list of contents. The table of contents is usually concise with titles of only the major sections. A table of contents is not a USPAP requirement but reflects the organization of the report and makes for easier reference.

Executive Summary

An executive summary is also not a USPAP requirement. It is, however, a professional courtesy as well as a convention. The executive summary, often referred to as the "Summary," can be placed in front or at the end of the report. This section is usually brief and provides the reader with an overview of the report. The summary usually includes

- the purpose and use of the report;
- the type or premise of value;
- the effective date of the appraisal;

- a discussion of other pertinent data regarding the appraisal, including its scope;
- an explanation of the appraisal's organization;
- possible solutions to particular appraisal questions (e.g., is there a market remedy?), and
- a recommendation and the reasoning behind the appraisal: if the appraiser is asked for a recommendation, it can be discussed in this section. (Typically, this is not the appraiser's responsibility but may be part of a consulting assignment.)

USPAP states that any time an appraiser renders an opinion of value, it is an appraisal. Therefore, a document, such as the Executive Summary, should stand on its own and cannot be misleading in the marketplace. For example, if the Executive Summary is separated from the report, will it be construed as the appraisal? If it contains an opinion of value, the answer is yes. Under normal reporting conventions, an Executive Summary is not the report but an overview and, if requested by the client, an appraiser recommendation. If the writer only wants to summarize the report data, the writer should clearly inform the reader of whatever facts are necessary to understand the meaning and limitations of the value. Other details can then be presented in the report.

Limiting Conditions

This is required by USPAP Standards Rule 8-2(a-viii, b-iii, and c-viii). A Statement of Limiting Conditions should be included with the initial engagement letter and in the final report. Limiting conditions are an integral part of the report. The conditions establish the framework for what the appraisal does and does not include. Appraisers may wish to obtain legal advice when preparing the limiting conditions. If the limiting conditions are exhaustive, the document may be redundant or even meaningless. Appraisers should be judicious editors of their statements and be prepared to tailor the limiting conditions to the assignment.

Definitions/Premise of Value

Reports should include definitions of the appraisal premise of value and other appraisal terminology. Some readers will not be familiar with appraisal terms. For example, the meaning that appraisers give to the term *fair market value* may differ from the client's understanding of the term. Remember that it is the appraiser's responsibility to select the proper definitions and perform the appraisal in accordance with the client's purpose. Definitions of value and appraisal terminology should be introduced early in the report. It may be appropriate to create a separate page for definitions. The writer may also define the concept when it is introduced in the report.

Identification and Description of the Property Appraised

This section provides a general description of the plant, process, or machinery appraised. It should serve as an overview and reference the asset inventory where more detailed information can be found.

Purpose of the Appraisal

This section of the report discusses the reason for preparing the appraisal: to determine a conclusion of value, under a premise of value, as of an effective date, for a specified asset or group of assets. The section should also address the intended use of the appraisal by the client or any other known users. Like the section identifying the appraised property, this area can be brief.

The Effective Date of the Appraisal

Everything in appraisal work is relative to the effective date of the appraisal. Therefore, the appraisal should always state the effective date of the valuation. It should be clearly stated so it is not confused with the transmittal date or the date of inspection. Normally the appraisal date is the last date that the appraiser made a physical inspection of the property, but this is not always the case. This delineation is important because the appraisal may apply to a date in the past or future. For example, a casualty insurance claim may require an appraisal as of the date of the loss. Conversely, a feasibility study of lease residual values may require a future effective date.

Methodology

The discussion of appraisal methods is one of the most important parts of the appraisal. This section should include a description of the extent of the process of collecting, confirming, and reporting data In this section, the appraiser will tell the reader exactly what contributed to the conclusion of value and how it was determined. Generally, the appraiser will discuss the property, how the appraised inventory was listed, and how the investigation was conducted.

Approaches to Value

Here, the appraiser will discuss the three approaches to value as required by USPAP Standards Rule 8-2). The appraiser will inform the client about the premise of value and why a particular appraisal method was used. If the appraiser decided to use more than one method of valuation—that is, both the cost approach and the sales comparison approach—the systematic logic to the approaches will be explained and concluded. Regardless of the approach the appraiser decides is most logical, the provisions of USPAP require the writer to consider all three approaches to value and explain why particular approaches were not used.

Summary

The Executive Summary was discussed earlier. However, the report itself should also contain a summary or conclusion. This summary is usually presented at the end of the report and reiterates the findings. This is a matter of personal preference, and not required by USPAP.

Explanation and Support of the Analysis of Highest and Best Use (Where Applicable)

This complies with reporting point number nine and USPAP Standard 8-2(a)(x), which states: "when the purpose of the assignment is any market value, describe the support and rationale for the appraiser's opinion of the highest and best use of the property." For machinery and technical valuation, a highest and best use analysis requirement will not always be either appropriate or applicable.

Normally, a highest and best use analysis involves examining the appraised property as a complete entity. Personal property, however, is usually composed of assets that are moveable. The highest and best use of a manufacturing line, for example, can change by pulling one machine out of the line and moving it elsewhere. If a machine or group of machines is located in a facility that complies with all regulations and if the plant is generating revenue, one assumes that the machinery is being put to its highest and best use. However, if the machinery appears to be usable but is idle because there is no demand for the product, or if the company has to be liquidated, one may assume that the equipment is not being used to its highest and best use. Highest and best use depends on the asset's ability to maximize business income.[22]

Other external factors influence highest and best use, such as with construction machinery (sometimes called "yellow goods" or "yellow iron") for which location and demand change constantly. Technological changes in high-tech electronic manufacturing are dynamic. These and other external influences emphasize the importance of the appraisal's date and time. The effective date of the appraisal can be a major factor in the determination of an asset's highest and best use.[23]

Appraisers need to furnish their own statement of highest and best use depending on the specifics of the appraisal. If the appraisal is of one item, such as a tractor/trailer rig, the highest and best use is when the rig is operating a certain number of hours per year. If a seasonal-product or holiday wrapping-paper plant is working only three months out of every twelve but uses single-purpose machinery, it may also be at its highest and best use. However, if only minor adjustments to the plant would make it operational year-round, or would enable it to work a full year, that plant has the potential to generate more revenue and achieve a higher and better use.[24]

In the technical valuation of a complete cement manufacturing plant, a refinery, a fluid milk plant, or other dedicated plants, the concept of highest and best use is an important consideration. Appraisers should understand that highest and best use is a matter of economic philosophy and is directly related to the manufacturer's regular product market and level of trade. Additional information on highest and best use can be found in most current real estate appraisal textbooks.

Appropriate Market

Where appropriate, an explanation and support of the analysis of the relevant market should be reported. For most MTS assets, the definition of value and intended use of the appraisal establish the appropriate market. For example, if the definition of value is *fair market value in continued use,* and the intention is to continue to use the assets as they exist, then the level of trade would probably be that in which an end user/buyer would acquire the assets for continued use. (Any discussion of highest and best use is also preempted by such a premise of value.) On the other hand, a liquidation premise of value might involve several different markets, such as a total entity auction versus a piece-by-piece auction.

Statement of the Appraiser's Disinterest, "Certification"

Both the ASA and USPAP standards require the Certification. The requirements and suggested contents were discussed earlier.

One of the most important disclosures in the Certification informs clients and prospective report readers of the appraiser's interest or lack of interest in the property appraised. Perhaps the appraiser is a dealer acting on behalf of the seller. If the appraiser is a broker and expects to participate in the sale of the property, this contemplated interest must be disclosed.

In addition, as previously stated, the ASA requires that a statement regarding its mandatory recertification program be included in the Certification:

> The American Society of Appraisers has a mandatory reaccreditation program for all of its Senior Members. 'I am' or 'I am not' in compliance with that program.

Qualifications of the Appraiser

Although not required by USPAP, appraisal reports often include the appraiser's credentials. This is especially true in an assignment for a new client or a report that will be used for court testimony.[25]

As experts, appraisers will detail their experience that qualifies them for the appraisal assignment. The statement of qualifications or "curriculum vitae" tells the client the appraiser's related work history, education, and appraisal background.[26]

Asset Inventory

This is what the appraisal is all about—the property appraised. Although the format is entirely up to the appraiser, the inventory is a list describing the assets or systems and their corresponding values. With the advent of personal computer software, often the inventory will be prepared in spreadsheet format. The appraiser should prepare the asset inventory in such a fashion consistent with the other sections of the report. For example, if the appraiser values a complete process line as a homogeneous unit, and the appraiser properly discusses this procedure in the report, then listing of the line as individual pieces with individual values will probably be misleading to the reader.[27]

Appendix or Addendum

Appraisers may use an appendix, if required, to gather illustrative and extraneous matter. When, and if, an appendix is used is a matter of reporting style. The appendix is a good place to put long examples, supporting material, and photos. Rarely, if ever, will an appraisal report require a cross-referenced index in the addenda. However, USPAP does require that all support data, as to values reported, be maintained in the appraiser's file.[28]

Signatures in Appraisal Reports and the Inclusion of Dissenting Opinions.

This topic is addressed in ASA Rule 8.8. Generally speaking, all parties that professionally contributed significantly to the appraisal must sign the report and must include a signed Certification in the report. The USPAP requirements are similar to those of the ASA, but explain in greater detail the requirements and responsibilities that are attached to report signatures.

The ASA requirement to include a dissenting opinion in the report is not a requirement of USPAP. Each professional appraiser operating under the auspices of the ASA should be aware of this extra ASA requirement.

Additional Information

USPAP Standard 8-2(b)(xi) states that the report must "state and explain any permitted departures from specific requirements of Standard 7, and the reason for excluding any of the usual valuation approaches."

It is important to remember that while departure from some of the requirements of USPAP is permitted, departure from others is not permitted. The report must clearly discuss any permitted departures that have been used in the appraisal.

Effective Writing Techniques

This chapter is not intended to instruct writers in proper grammar or encroach on a writer's style. Report writers concerned with effective writing techniques can consult any of several writing guides available (see "Additional Reading" at the end of this chapter).

To reiterate, there is **a beginning, a middle,** and **an end** to every report. An effective report writer knows when to begin, consider, and present the supporting data, and when to end the report. An effective report writer is brief, clear, and factual.

Clarity

Remember that USPAP Standard 8-1 requires clarity and that the report must not be misleading. The appraiser can assist the reader here by writing clearly and succinctly. Also, avoid abbreviations and any appraisal jargon. Avoid terms such as "as you know" or "as we discussed," as every reader may not be privy to these data.

Transitions

Transitions, not always self-evident, are words or phrases that bridge paragraphs. An effective writer can use transitions to help make the report easier to follow.[29]

Spelling/Typos

A few spelling errors or typos will quickly erase the professional intentions of the appraiser. Such mistakes can demonstrate that the writer has not taken the time to carefully review the report and might prompt the reader to believe there may be other mistakes as well. With current computer spell-check programs, there is no excuse for poor spelling.[30]

Math

Mathematical errors in appraisals are not tolerable. Most appraisers now use electronic spreadsheets that calculate all of the totals. The computer does not distinguish between good or bad input, however, and appraisers should always check their data entry. Appraisers can develop a method of sampling long calculations to ensure that no errors or faulty decimal extensions are made.[31]

Graphics

Graphics are useful for demonstrating the relationship of manufacturing lines or special installations, charting values by account, and showing totals by location. Often, they can convey more than a photo or other exhibit. Graphics are increasingly accessible using scanners, desktop publishing, word processing, and spreadsheet software.[32]

Report Style

The reality of the appraisal occupation is that most clients want the work done professionally but also quickly and economically. To be competitive, appraisers may need to do something less than a full narrative or "self-contained" report, but these "summary" and "restricted" appraisal reports must comply with USPAP and ASA (if the appraiser is an ASA member) requirements.

Narrative Report

The narrative report essentially contains the information required to substantiate the findings, although some data may be located in the appraiser's files and referenced in the report. The styles and layouts may differ, yet still conform to USPAP and ASA requirements. Parts of the appraisal that are used repetitively, that are "boilerplate," can be stored in a word processing document in the writer's computer so that future reports can be created more easily. For example, the transmittal letter, title page, assumptions and limiting conditions, definitions, certification, and qualifications may be the same or quite similar from report to report. Once the report format is stored in a computer file, the subsequent writing becomes less tedious.

Form Report

For smaller, less complicated appraisal reports, a "form report" can be developed. A form report contains a great deal of boilerplate with an area for the inventory and the pertinent narrative. Form reports have distinct headings and clearly identified data or graphic sections. Form reports work best for repetitive assignments like construction equipment, trucks, automobiles, and even some kinds of manufacturing facilities such as machine shops. A form report works especially well if the appraiser is also a machinery dealer and works exclusively with certain types of equipment. There are no set prescriptions for form reports, and the design is a matter of personal preference. They must, however, still meet *all* of the USPAP requirements.[33]

Oral Reports

Clients and potential clients often ask appraisers for "quick-and-dirty" estimates of value. The quickest road to trouble is to render an "unconsidered opinion." The appraiser must adhere to USPAP Standards Rule 8-1, as listed below, when considering an oral report.

> **Standards Rule 8-1**
>
> Each *written* or *oral* personal property appraisal report must
>
> - clearly and accurately set forth the appraisal in a manner that will not be misleading;
> - contain sufficient information to enable the intended users of the appraisal to understand the report properly;
> - clearly and accurately disclose any extraordinary assumption, hypothetical condition, or limiting condition that directly affects the appraisal and indicate its impact on value.

Appraisers who intend to comply with USPAP and ASA requirements are responsible for their opinions.[34] Oral reports carry the same responsibilities as a written report. It is advisable to follow up any verbally transmitted opinion with a written confirmation. Any opinion of value rendered must meet USPAP requirements!

Key Points

- The result of an appraisal assignment is a report.
- A report is a document that can be relied on as factual and can be used as a tool to assist clients in making decisions.
- The ASA and USPAP have standardized requirements for proper appraisal practice. Some of these requirements will change as USPAP evolves.
- It is the responsibility of appraisers to keep current with changes in appraisal reporting requirements.
- There are conventions unique to appraisal reports that are not normally found in other types of consulting.
- Writing appraisal reports will not be difficult if the appraiser remembers the client, the use, and the requirements of proper appraisal practice.

Additional Reading

Anderson, Paul V. Technical Writing. 2nd ed. Orlando, FL: Harcourt Brace Jovanovich College Publishers, 1989.

Bowers, Donald E., and James Lee Young, eds. *The Professional Writer's Guide.* Aurora, CO: National Writers Association, 1990.

Bunnin, Brad, and Peter Beren. *The Writer's Legal Companion: How to Deal Successfully with Copyrights, Contracts, Libel, Taxes, Agents and Publishers, Legal Relationships and Marketing Strategies.* New York: Addison-Wesley, 1988.

Evans, Glen, ed. *The Complete Guide To Writing Non-Fiction.* New York: Harper & Row, 1983.

Haines, Howard. "Writing Better Reports," *American Society of Appraisers, Appraisal and Valuation Manual, 1961–1963, Volume 7.* Washington, DC: American Society of Appraisers, 1964.

Himstreet, William C. *Communicating the Appraisal: The Narrative Report.* Chicago: American Institute of Real Estate Appraisers, 1988.

Himstreet, William C. *Writing Appraisal Reports.* Chicago: American Institute of Real Estate Appraisers, 1974.

The Chicago Manual of Style. 13th ed. Chicago: The University of Chicago Press, 1982.

USPAP 98, *Uniform Standards of Professional Appraisal Practice, 1998 Edition.* Washington, DC: Appraisal Standards Board, The Appraisal Foundation, 1998.

Zinser, William. *On Writing Well: An Informal Guide To Writing Non-Fiction.* 4th ed. New York: Harper Perennial, 1990.

Notes

[1] "Writing Better Reports," by Howard Haines, ASA. American Society of Appraisers Appraisal and Valuation Manual 1961-1963, Volume 7.

[2] Alan C. Iannacito, *Bolt by Bolt.* Manuscript, Golden, Colo.

[3] The reader is advised that this chapter discusses only report format and style. It is the appraiser's responsibility to comply with the Uniform Standards of Professional Appraisal Practice (USPAP). USPAP reporting requirements may change as dialogue and meaning are reviewed or revised by subsequent USPAP advisory boards. The information in this chapter is a general description of an evolving process and should not be relied on as an authoritative interpretation of USPAP, ASA, or other appraisal society requirements.

[4] Iannacito, "Bolt by Bolt."

[5] All discussion of USPAP is based on the January 2000 edition; obviously, this chapter cannot address changes that may be made to USPAP after this edition, and the reader is advised to consult the latest edition of USPAP.

[6] The following text is quoted verbatim from Standards Rule 8-1 in USPAP 2000: Uniform Standards of Professional Appraisal Practice, 2000 Edition (Washington, D.C.: Appraisal Foundation, 2000), 54.

[7] Ibid

[8] Ibid., pp.55–57.

[9] Ibid., p.120.

[10] This is a minimum list from USPAP 2000, p.61. The appraiser should have access to current USPAP documents that comment on the standards and discuss "departure" provisions.

[11] American Society of Appraisers, Washington, DC, authorized June 30, 1968, revised January 1994.

[12] USPAP 98, 2000, p.4.

[13] Ibid., pp. 54–60.

[14] Iannacito, "Bolt by Bolt."

[15] Ibid.

[16] Ibid.

[17] Ibid.

[18] Ibid.

[19] Ibid.

[20] Ibid.

[21] USPAP 98, *Uniform Standards of Professional Appraisal Practice, 1998 Edition* (Appraisal Standards Board, The Appraisal Foundation, Washington, D.C. 20090-6734), pp. 50–51.

[22] Iannacito, "Bolt by Bolt."

[23] Ibid.

[24] Ibid.

[25] Ibid.

[26] Ibid.

[27] Ibid.

[28] Ibid.

[29] Ibid.

[30] Ibid.

[31] Ibid.

[32] Ibid.

[33] Ibid.

[34] Ibid.

7

Ethics

Objectives:

1. Define personal and business ethics.
2. Define appraisal ethics.
3. Explain the Uniform Standards of Professional Appraisal Practice (USPAP).
4. Explain the American Society of Appraisers' Principles of Appraisal Practice and Code of Ethics (ASAPAP).

This chapter begins with a discussion of personal and business ethics but focuses on the ethical procedures and standards that professional appraisers are expected to follow in their appraisal assignments; it also provides guidance on the framework within which professional appraisers can maintain their fiduciary responsibilities. The appraiser's fiduciary responsibility emanates from a relationship of trust, a relationship that is not necessarily imposed by law. This chapter is not a philosophical presentation of ethical standards. It is a listing of the specialized ethics that the professional appraiser is expected to follow.

The basic ethics required by the *Uniform Standards of Professional Appraisal Practice* (USPAP) as promulgated by the Appraisal Standards Board (ASB) of the Appraisal Foundation are covered here along with the American Society of Appraisers' *3 Principles of Appraisal Practice and Code of Ethics* (ASAPAP). Professional appraisers conduct their practices in compliance with USPAP. The ASA is one of the original sponsoring organizations of the Appraisal Foundation and has adopted USPAP. All ASA members must conform to USPAP.[1]

Personal and Business Ethics

Webster's New World Dictionary [second college edition] defines ethics as "the system or code of morals of a particular person, religion, group, profession, etc." Morals are defined in *Webster's* as "principles, standards, or habits with respect to right or wrong in conduct." Ethics and morals are used synonymously, but there is a difference: in general, *morality* refers to the rules and standards of conduct of a specific society at a certain time and place, whereas *ethics* usually refers to the rules and norms of specific kinds of conduct or the codes of conduct for specialized groups.

In everyday life rules of conduct dictate what is right or wrong in many situations. The same relationship does not always legally apply. What is legal is not necessarily ethical, and what is unlawful is not necessarily unethical. For example, driving 65 miles per hour where the speed limit is 55 miles per hour is not unethical, but it is considered illegal. Although it is not illegal in some jurisdictions for a psychiatrist to initiate a social relationship with a patient, it is certainly unethical.

Since business activity is controlled by many laws, it could be argued that anything that is legal is necessarily ethical. The profit motive in business, however, is a strong stimulus that adds to the ethical confusion and dilemma. The Ethics Resource Center in Washington, D.C., conducted a survey in 1994 of approximately 4,000 individuals from various companies. Many of those surveyed said that when their company had to make a choice between doing what was right and making a profit, the profit choice was made most of the time. Nearly one-third of the employees surveyed sometimes felt pressured to engage in misconduct and at least occasionally ignored ethical standards to meet business objectives. For the professional appraiser, this compromise is not an alternative; ethical standards are paramount.

The hypothetical appraisal case that follows illustrates some of the differences and the specialized nature of appraisal ethics. The remainder of the chapter deals with ethical issues.

Hypothetical Appraisal Case

> An appraiser, knowing the fee that a competing appraiser quoted, reduced his fee quotation to a longtime client in an attempt to underbid the competing appraiser. Nevertheless, the appraiser still lost the assignment. The client was impressed with the names of the clients furnished by the other appraiser and awarded her the assignment. The unsuccessful appraiser knew that the successful appraiser had no experience with the type of equipment to be included in the appraisal. The appraiser who completed the assignment later purchased some of the equipment that she had appraised for the client.

What action, if any, should the unsuccessful appraiser take and what issues or problems may be involved in this hypothetical appraisal case? Consider this appraisal case from an individual perspective and from a business perspective. The issues involved are

- reduction of appraisal fees,
- disclosure of previous appraisal clients,
- inexperience,
- criticism of another appraiser, and
- appraisal and subsequent purchase of the same equipment.

Competition is a way of life in our society, and a price discount is not an issue in personal or business ethics. Disclosure of past successes is a competitive tactic and is not considered unethical in business; of course, the appraiser should be careful that any disclosure does not violate confidentiality requirements imposed by USPAP or the client. The appraiser's inexperience could be a problem in a continuing business relationship but is not necessarily an ethical issue. Criticism of another appraiser could be an ethical problem for some individuals. The apparent conflict of interest in the last issue seems to be the only real problem from an individual and possibly from a business ethical perspective. Nevertheless, all of these issues are of ethical concern for the professional appraiser.

Appraisal Ethics

The glossary of this text defines an appraisal as "an unbiased opinion of value of an identified property based upon market research; assemblage of pertinent data; application of appropriate analytical techniques; and knowledge, experience, and professional judgment." The two important ethical terms in the definition are an *unbiased opinion* and *professional judgment.*

An appraiser doing an appraisal cannot be an advocate. An advocate is biased. *Webster's* defines an advocate as "a person who pleads another's cause; specifically, a lawyer." Note the apparent redundancy of an "appraiser doing an appraisal." (A USPAP exception on advocacy that explains this redundancy is discussed later in this chapter.) Although an appraiser is hired and paid by the client, the appraiser cannot be an advocate of the client's cause. An appraiser is a professional who is duty-bound to submit an unbiased opinion.

This duty is the fiduciary responsibility of the appraiser to both the client and to third parties. *Fiduciary* means based on firm faith and connotes a relationship of trust. The fiduciary responsibility extends to any third party who may rely on the appraisal report. This unbiased duty is essential to maintain appraising as a respected profession. An appraiser cannot accept an assignment with special conditions dictated by the client because such conditions might bias the appraisal. This proscription may seem to contradict the adage that the customer is always right, but the rationale behind it becomes obvious when the purposes of the opinion are considered. For example, MTS appraisals are commonly used for financial purposes; in such cases, lending institutions are third parties to which the appraiser has a fiduciary responsibility. Appraisers in some federally regulated transactions, as addressed in USPAP advisory opinion AO-10, must be directly engaged by the lender.

The appraiser's fiduciary responsibility also extends to the public. This could apply to depositors of a financial institution that uses appraisals for loan purposes or to taxpayers in an eminent domain acquisition. The public is also a third party that has the right to rely on the validity and objectivity of the appraisal report.

The appraiser develops professional judgment through education, training, research, study, practice, and experience. An appraiser must have the expertise and knowledge to competently perform the assignment. This is addressed further under the sections on USPAP and ASAPAP.

As a professional appraiser, consider the issues raised in the hypothetical appraisal case. The first two issues, "reduction of appraisal fees" and "disclosure of previous appraisal clients," are discussed in the sections on USPAP and ASA that follow. The "inexperience" issue may be construed as an infraction of fiduciary responsibility under general appraisal ethics. The "criticism of another appraiser" issue is questionable in a respected profession. The "appraisal and subsequent purchase of the appraised equipment" issue is definitely an ethical problem for any appraiser because it raises the possibility of a self-serving appraisal opinion. It is also questionable whether appraisers can render truly unbiased opinions about value if they are also prospective purchasers. This problem, discussed in subsequent sections, is handled differently by USPAP and ASA. Many financial institutions do not permit the appraiser to be a prospective buyer even if, or especially if, that future interest is revealed. Appraisers who also wish to purchase the equipment may have to forgo the appraisal assignment so as not to disqualify themselves as prospective purchasers.

The appraiser does not have the luxury of ethical compromise. Professional appraisers must be willing to place personal interest second to the interest of the public whom they serve. Ethical considerations must supersede economic considerations.

Uniform Standards of Professional Appraisal Practice (USPAP)

The major appraisal societies formed an ad hoc committee on uniform standards, and that committee originally developed USPAP. The ASA was the only society involved that comprises appraisal disciplines other than real property. In the next chapter the ad hoc committee, the Appraisal Foundation, the Appraiser Qualifications Board (AQB), and the Appraisal Standards Board (ASB) are discussed. Appraisers and users of appraisal services should have a copy of the current edition of USPAP, a living document with biannual updates and annual editions. It is available from the Appraisal Foundation (address listed in "Additional Reading" section).

The ASB continually develops, amends, interprets, and publishes USPAP. Neither the ASB nor the Appraisal Foundation is an enforcer of USPAP. Rather, USPAP is enforced by peer review; complaints are directed to the respective societies. A court of law is many times the ultimate ethics enforcer for the appraisal professional. The establishment and work of the Appraisal Foundation have contributed to the evolution of appraisal from a trade into an accepted profession.

Standards 7 and 8 of USPAP address personal property appraisals. Other parts of USPAP also apply to personal property appraisers. USPAP begins with and includes the preamble. All of the USPAP preamble, all of Standards 7 and 8, and some of the USPAP statements and advisory opinions apply to the personal property appraiser.

The word "misleading" is used again and again throughout USPAP. The first sentence of Standard 8 states that an appraiser "must communicate each analysis, opinion and conclusion in a manner that is not misleading." This concept continues to appear throughout the ethics provision and the standards.

The ethics provision of the preamble initially reminds appraisers of their inherent fiduciary responsibility. The conduct section further states that it is unethical to use or communicate a misleading or fraudulent appraisal. An assignment based upon a hypothetical condition serves as an example of required disclosures. The fiduciary responsibility, a professional relationship of trust, dictates an unbiased opinion.

The section on management addresses appraisal fees. It states that contingent fees are unethical and that an appraisal fee cannot be predicated on the value reported. Even a direction of value favoring the client cannot be reported. Many appraisal firms insist upon payment in full before the release of any valuation conclusions. This procedure ensures that the client has no influence on the values reported. The management section also states that the payment of any fee or receipt of a thing of value for an appraisal assignment must be disclosed.

Consider the "thing of value" issue in the preceding hypothetical case. If the appraiser reduces a fee with the intention of obtaining the assignment, the appraiser is offering a thing of value. The comment in the management section clearly states that disclosure of fees, commissions, or things of value connected to obtaining an assignment should appear in both the certification and the letter of transmittal. The USPAP instructors' guide published by the ASB further explains this concept thus: "If an assignment is obtained by lowering the appraiser's normal fees through negotiation or through competitive bidding, this is providing a 'thing of value' that should be disclosed in the appraisal report's Certification and Transmittal Letter...." Competition is an accepted fact in our society, but the professional appraiser has further obligations in the competitive arena. It is conceivable that the initial negotiation may have some relevance to a third party who is relying on the appraisal report. The "thing of value" issue does not apply to a fee adjustment in response to a change in the scope of a project.

The neutrality of a professional appraiser is a requirement of all major appraisal societies. An appraiser cannot be an advocate and must report an unbiased opinion. Yet, USPAP states an exception to this age-old principle in the management section of the ethics provision. Remember the apparent redundancy of an "appraiser doing an appraisal." The contingent compensation and the advocacy restriction do not apply in consulting assignments in which the appraiser would *not reasonably be perceived* as acting in a disinterested manner or "...performing a service that requires impartiality." The required certification for consulting, standards rule 5-3, is the only certification that provides an optional response: "My compensation is not (or is) contingent...."

The exception on contingent compensation and advocacy does not apply to all consulting services. Consulting is a part of appraisal practice. The glossary defines an appraisal practice (USPAP) as "the work or services performed by appraisers, defined by three terms in the Uniform Standards: appraisal, review, and consulting." In a consulting assignment, all ethical considerations apply except when it would be obvious to any third party that the appraiser is not unbiased. The fact that the appraiser is not acting in a disinterested manner must be properly disclosed in the report. This exception is further covered in standard 4, performing a consulting service, and standard 5, reporting a consulting service. Although standards 4 and 5 apply to real property, it is often interpreted as being applicable to personal property also.

This discussion and USPAP references illustrate that all of the preamble and ethics provision, specifically standards 4 and 5, apply to all disciplines including personal property. The explanatory comments are also an integral part of USPAP. The USPAP statements and advisory opinions also apply to all professional appraisers. The statements have the full force and effect of the standards rules, whereas the advisory opinions are meant to provide useful guidance. Although a section of USPAP may address the real estate or real property appraiser, as do standards 4 and 5, this does not exempt the personal property appraiser from the requirement. The professional personal property appraiser, and specifically all members of the American Society of Appraisers, must conform to all applicable parts of USPAP, not just standards 7 and 8, which address personal property.

The "disclosure of previous appraisal clients" issue in the hypothetical appraisal case involves the confidentiality responsibility of an appraiser. USPAP states that "an appraiser must protect the confidential nature of the appraiser-client relationship." Factual data or results of an assignment cannot be disclosed except to someone authorized by the client, as required by law, or to a peer review committee. Confidentiality is fundamental to the appraiser-client relationship. In fact, appraisers are sometimes required to sign confidentiality agreements. Revealing the names of past clients could be a breach of confidentiality.

The USPAP statement on appraisal standards (SMT-5) is an additional explanation of the confidentiality rule of the ethics provision. This statement acknowledges that the obligation is neither absolute nor clearly understood and further explains that the appraiser-client relationship is not as strong or inviolate as the attorney-client relationship. The USPAP necessity of revealing information to a peer review committee could conflict with, but be superior to, an all-encompassing appraiser-client confidentiality agreement. In addition, some of the market data obtained in an assignment may be useful in future assignments. To hold all data obtained as confidential may be unreasonable. Clearly, the result of an assignment is confidential as well as all information so specified by the client. The client, however, must be informed of the appraiser's responsibility to conform to all USPAP requirements. Any confidentiality agreement must be worded so that the appraiser can conform.

The "inexperience issue" in the hypothetical appraisal case is addressed in the competency provision of USPAP. An appraiser must have the knowledge and experience to perform competently. It is important for the professional appraiser to participate regularly in continuing education activities. All ASA-designated appraisers must successfully participate in the Society's mandatory reaccreditation program.

If an unfamiliar assignment is encountered, it is unethical for an appraiser to accept or continue without applying all the following three steps:

1. Disclose the lack of knowledge and/or experience to the client before accepting the assignment.
2. Take all steps necessary or appropriate to complete the assignment competently.
3. Describe in the report any lack of knowledge or experience and the steps taken to complete the assignment competently.

The lack of training, skill, and experience to perform competently may be offset by study and research. If that is not possible or feasible, an association with another appraiser or the retention of an expert may be considered. The client must be made aware of the lack of expertise and concur with the steps taken. The unsuccessful appraiser in the hypothetical appraisal case has no way of knowing if the successful appraiser handled her alleged inexperience in the assignment properly. The competitor may have disclosed her inexperience to the client and correctly resolved the inexperience problem.

Appraisers should be conservative in estimating their own professional competency. Appraisers are rarely impartial judges of their own competency, which is often judged by others in hearing rooms or courtrooms. *A wise rule is "If you have to ask whether you need help, you probably do."*

The ethics and competency provisions permeate USPAP and apply to all appraisal disciplines. These provisions include the standard rules, the comments, the statements on appraisal standards, and the advisory opinions. This importance of these provisions is underscored by the certification required to be a part of all appraisal reports for all appraisal disciplines.

USPAP does not directly address the "criticism of another appraiser" issue presented in the hypothetical case, but USPAP does continually refer to appraisers as professionals. The ASA does address this issue and considers it unprofessional and unethical to injure the professional reputation of another appraiser. This concept is discussed in the next section on the principles of appraisal practice and the code of ethics (ASA).

The last issue in the hypothetical case, "appraisal and subsequent purchase of the same equipment," is addressed by USPAP in the required certifications (Standards Rules 2-3, 3-2, 5-3, 6-8, 8-3, 10-3). The certifications section requires that the appraiser must sign a statement to the effect that "I have no (or the specified) present or prospective interest..." in the equipment or the parties involved. The "apparent" conflict of interest is permissible according to USPAP if it is disclosed in the certification. Reflection of an appraiser's fiduciary responsibility is required to understand this apparent conflict. Third parties, who may subsequently rely upon the appraisal, must be made aware of the appraiser's interest in the property so that they can assess for themselves the possible effect on the results.

The conduct section of the ethics provision requires that appraisers "...perform assignments with impartiality, objectivity, and independence and without accommodation of personal interests." Therefore, appraisers should strive to ensure that their other activities, such as also being an equipment broker or dealer, do not influence, or even appear to influence, their objectivity as professional appraisers.

Principles of Appraisal Practice and Code of Ethics (ASAPAP)

Ethical practices and the conduct required of ASA members are defined in ASAPAP. In some respects, these requirements are more stringent than the basic ethics required by USPAP. They define for ASA members and users of appraisal services what competent and ethical practice is and characterize certain practices as being unethical and unprofessional.

Regarding the issue of "criticism of another appraiser," these standards state that the ASA appraiser is obligated to protect the professional reputation of all appraisers. Unethical conduct by an appraiser reflects not only upon that individual but also upon all appraisers and all professional appraisal societies. Section 5.1 of ASAPAP states that "...it is unethical for an appraiser to injure, or attempt to injure, by false or malicious statements or by innuendo the professional reputation or prospects of any appraiser." An ASA member has an obligation to report to the Society any violation of ASAPAP. In the hypothetical case, the appraiser cannot allege the incompetence of the competitor to the client but must report any ethical violation to the Society.

In addition, ASA deals with the obligation to give competent service (the "inexperience issue") in much the same way that USPAP does. The appraiser must fully acquaint the client with any limitations and associate with another appraiser who has the required qualifications. Both ASA and USPAP extend the competency principle to advertising.

False, misleading, or exaggerated advertising of solicitation for assignments is unethical according to USPAP. Section 7.7 of ASAPAP states that it is unethical to

(a) misrepresent in any way one's connection or affiliation with ASA or any other organization;

(b) misrepresent one's background, education, training or expertise;

(c) misrepresent services available or an appraiser's prior or current service to any client, or identify any client without the express written permission of such client to be identified in advertising;

(d) represent, guarantee or imply that a particular valuation or estimate of value or result of an engagement will be tailored or adjusted to any particular use or conclusion other than that an appraisal will be based upon an honest and accurate adherence to the Principles of Appraisal Practice.[2]

Concerning the "disclosure of previous clients" issue, the ASA is more direct and restrictive on the appraiser's confidentiality obligation than is USPAP. Section 4.1 of ASAPAP states that "the fact that an appraiser has been employed to make an appraisal is a confidential matter." The client may have valid reasons for keeping the fact of employment of an appraiser as confidential. ASA continues by stating that "...it is improper for the appraiser to disclose the fact of his engagement, unless the client approves of the disclosure or clearly has no interest in keeping the fact of the engagement confidential...."

Although USPAP allows for the possible necessity of revealing information to peer review committees, the ASA does not. Section 4.1 also states that "in the absence of an express agreement to the contrary, the identifiable contents of an appraisal report are the property of the appraiser's client or employer and, ethically, cannot be submitted to any professional society as evidence of professional qualifications, and cannot be published in any identifiable form without the client's or employer's consent."

In its initial description of the objectives of appraisal work, ASA states that all the principles of appraisal ethics stem from the central fact that the result is objective and unrelated to the wishes of the client. This could not be stated more directly. An appraiser must submit an unbiased opinion and not be an advocate in any way.

On advocacy, section 7.5 of ASAPAP states that "if an appraiser...suppresses or minimizes any facts, data, or opinions, which, if fully stated, might militate against the accomplishment of his client's objective or, if he adds any irrelevant data or unwarranted favorable opinions or places an improper emphasis on any relevant facts for the purposes of aiding his client in accomplishing his objective, he is, in the opinion of the Society, an advocate." Advocacy adversely affects the trust and confidence in professional appraisal practice. It is in direct conflict with an appraiser's fiduciary responsibility. It is unethical and unprofessional.

Contingent fees and percentage fees are also unethical and unprofessional. If an appraiser were to accept such a fee then "...anyone considering using the results of the appraiser's undertaking might well suspect that these results were biased and self-serving and therefore, invalid." As a corollary to the above principle relative to contingent fees, section 7.1 declares that it is unethical and unprofessional for an appraiser

(a) to contract for or accept compensation for appraisal services in the form of a commission, rebate, division of brokerage commissions, or any similar forms and

(b) to receive or pay finder's or referral fee.[3]

The "appraisal and subsequent purchase of the same equipment" issue in the hypothetical case is addressed directly by the

ASA in section 7.3 under the rubric of disinterested appraisals. The user of an appraisal in which the appraiser has a present or contemplated future interest "...might well suspect that the report was biased and self-serving and, therefore, that the findings were invalid." The ASA declares this unethical and unprofessional with the following exception: A full disclosure must be made by the appraiser before the fact, and the appraisal report must disclose the nature and extent of the appraiser's interest. This disclosure correlates with the USPAP-required certification that the appraiser has no, or a specified, interest. USPAP does not directly address this issue except in the certifications.

The "reduction of appraisal fees" issue in the hypothetical case is not directly addressed in ASAPAP. A connection might be argued under the concept of contingency and percentage fees or disinterested appraisals. Such a discussion, however, would be merely academic because the ASA appraiser must adhere to USPAP as well. According to USPAP, such a situation must be disclosed in the report.

The appraisal report section of the ASA Principles that pertains to USPAP reads as follows:

8.6 Appraiser's Responsibility to Communicate Each Analysis, Opinion and Conclusion in a Manner That Is Not Misleading.

> The appraiser should state in each report "I hereby certify that, to the best of my knowledge and belief, the statements of fact contained in this report are true and correct, and this report has been prepared in conformity with the Uniform Standards of Professional Appraisal Practice of the Appraisal Foundation and the Principles of Appraisal Practice and Code of Ethics of the American Society of Appraisers."

The professional appraiser, and specifically the ASA appraiser, must conform to all aspects of USPAP and the requirements of his or her respective society.

The ethics section of ASAPAP addresses signatures in appraisal reports. Three sample cases are cited in regard to the user of the appraisal. The user is entitled to assume that the signer of the report is responsible for all findings. In a joint report, both signers are jointly and equally responsible, and the user has the right to know of any dissenting opinions. If two or more appraisers are engaged to make independent appraisals, the user has the right to expect that the opinions were reached independently.

Signers of reports are dealt with by USPAP in several different sections. The required certification must identify signers who have inspected the property or provided significant professional assistance. The inspection aspect is illustrated and clarified in advisory opinion AO-2. This advisory opinion addresses real property, but the references to "drive-by" and "walk-through" inspections can apply to personal property. If the inspection was of some special nature, it should be so noted in the report. Advisory opinion AO-5 illustrates and clarifies the assistance in the appraisal. As in most professions, the principal is responsible for the work of assistants; even if the assistant does not sign the report, his or her assistance must be acknowledged.

Standards rule 8-5 on the reporting of a personal property appraisal states that "an appraiser who signs a personal property appraisal report prepared by another, even under the label of 'review appraiser,' must accept full responsibility for the contents of the report." Standards rule 3 addresses review appraisers, and this concept is further illustrated and clarified in an advisory opinion (AO-6).

A mandatory recertification statement is required in all appraisal reports by ASA as detailed in section 8.7. All designated members should state in each report: "The American Society of Appraisers has a mandatory reaccreditation program for all of its Senior and Accredited members. 'I am' or 'I am not' in compliance with that program."

ASA has three professional grades of membership: accredited member (AM), accredited senior appraiser (ASA), and fellow (FASA). It is unethical for anyone to claim or imply holding a higher degree of membership than has been attained.

As stated in section 7.9, ASA takes disciplinary action against members for violations that may be in the following categories:

- deviations from good appraisal practice;
- failure to fulfill obligations and responsibilities;
- unprofessional conduct;
- unethical conduct;
- conviction of felonies and misdemeanors, especially involving honesty or veracity; and
- any unlawful, illegal, or immoral conduct that would bring disrepute to the appraisal profession or to ASA.

Key Points

- All members of the American Society of Appraisers must comply with the Society's *Principles of Appraisal Practice and Code of Ethics*, as well as the *Uniform Standards of Professional Appraisal Practice* as promulgated by the Appraisal Foundation.

- An appraisal is an unbiased opinion of value. All the principles of appraisal ethics stem from the central fact that the result is objective and unrelated to the wishes of the client or future actions of the appraiser. If an appraiser has any interest in the property or parties involved, it must be properly disclosed in the report.

- *Reduction of appraisal fees:* In the hypothetical appraisal case, the appraiser proposed a lower-than-normal fee but was not awarded the assignment. It is unethical for an appraiser to reduce a fee quote below a known competitive fee quote after an initial fee quote has been submitted. This type of reduction in fee is considered a "thing of value" by the USPAP. This reduction could influence a third party who has the right to know of anything that may materially affect the appraisal. The appraiser has a fiduciary responsibility toward the possible third party and must, therefore, disclose such a situation in the report.

- *Disclosure of a previous appraisal:* In the hypothetical appraisal case, the appraiser won the job based on the use of previous clients as references. For the professional appraiser, this violates the confidentiality that is vital to the appraiser-client professional relationship. This is decisively addressed by ASA: "The fact that an appraiser has been employed to make an appraisal is a confidential matter." Confidentiality is also a rule of the ethics provision of USPAP and is further addressed in the statement on appraisal standards number SMT-5. USPAP admits that this obligation is neither absolute nor clearly understood. All information so specified by the client is most certainly confidential under normal circumstances. Disclosure of former clients as references is allowed if permission to do so has been granted.

- *Inexperience:* In the hypothetical appraisal case, the losing appraiser believed that the winning appraiser had no direct experience. This presents a major problem for the professional appraiser. Even with the approval of the client, the appraiser has a fiduciary responsibility to intended third parties. Both ASA and USPAP address the appraiser's obligation to give competent service. The appraiser must fully acquaint the client with the limitation, associate with another to be able to competently perform the assignment, and disclose the steps taken in the appraisal report. These steps are outlined in USPAP's competency provision. The disclosure is necessary to fulfill the fiduciary responsibility.

- *Criticism of another appraiser:* This issue, as presented in the hypothetical appraisal case, deals with the question of whether the winning appraiser's potentially unethical behavior should be reported. From an appraisal standpoint, unethical conduct by one appraiser damages all in the profession; but one appraiser should not openly criticize another appraiser. This in itself is unprofessional. Violations of conduct should be filed with the appropriate society. ASA members have an obligation to report violations.

- *Appraisal and subsequent purchase of the same equipment:* In the hypothetical appraisal case, the winning appraiser subsequently purchased some of the equipment that was the subject of the appraisal. Both ASA and USPAP recognize that this may be acceptable in some instances. If an appraiser does have an interest in the subject property, that interest must be fully disclosed in the report. This, too, is an appraiser's fiduciary responsibility toward any intended third parties.

- *Advocacy*: An appraiser cannot be an advocate by slanting an appraisal to accommodate the desires and interests of the client with the exception of consulting assignments in which the appraiser would *not reasonably be perceived* as acting disinterestedly. Appraisals should be meaningful to the client but not *misleading* in the marketplace. An appraiser also has a fiduciary obligation. This is a relationship of trust toward the client and third parties who have the right to rely on the validity and objectivity of the appraisal. This trust also extends to confidentiality of work performed. The results of an assignment and all else so specified by the client are confidential.

- *Contingency or percentage fees:* Contingency or percentage fees charged by an appraiser for appraisal services are unethical. In certain consulting assignments in which advocacy is obvious, this admonition does not apply. An appraiser must have the knowledge and experience to perform competently. The client must be informed of any limitation; a remedy or association with a qualified appraiser must be arranged; and full disclosure must be in the report.

Additional Reading

The Appraisal Foundation. *A Guide for Instructors Teaching USPAP, Appraisal.* Washington, D.C.: Standards Board of the Appraisal Foundation. Updated annually.

The Appraisal Foundation. *Uniform Standards of Professional Appraisal Practice*. Washington, D.C.: Standards Board of the Appraisal Foundation (1029 Vermont Avenue NW, Suite 900, Washington, D.C. 20005, (202) 347-7722). Living document with annual editions.

Boatright, John R. *Ethics and the Conduct of Business*. Englewood Cliffs, N.J.: Prentice Hall, 1993.

Goodell, Rebecca. *Ethics in American Business: Policies, Programs and Perceptions*. Washington, D.C.: Ethics Resource Center (1120 G Street, Suite 200, Washington, D.C. 20005, (202) 737-2258), 1994.

Principles of Appraisal Practice and Code of Ethics. Washington, D.C.: American Society of Appraisers, 1994.

Rand, Ayn, and Leonard Peikoff, eds. *The Voice of Reason*. New York: Meridian, 1989.

Notes

[1] The enforcement, or use of the USPAP is described in the Preamble (USPAP 2000, page 1) as follows:

> These Standards are for appraisers and users of appraisal services. To maintain a high level of professional practice, appraisers observe these Standards. However, these Standards do not in themselves establish which individuals or assignments must comply; neither The Appraisal Foundation nor its Appraisal Standards Board is a government entity with the power to make, judge, or enforce law. Individuals comply with these Standards either by choice or by requirement placed upon them, or upon the service they provide, by law, regulation, or agreement with the client or intended users to comply.

[2] *Principles of Appraisal Practice and Code of Ethics* (Washington, D.C.: American Society of Appraisers, 1994), p.9.

[3] Ibid., p.7.

8

Leasing

Objectives:

1. Define important leasing terms.
2. Summarize the history of leasing.
3. Explain the four basic types of lease structures.
4. Discuss the role of appraisals in leasing.
5. Introduce residual forecasting.

Equipment leasing, as we know it today, is primarily a product of various tax and accounting rules and practices that have evolved over the years. Leasing is a growing area that sometimes requires professional appraisals to assist lessors with confirming lease structures and the value of assets being leased. The purpose of this chapter is to provide a basic understanding of equipment leasing; its appraisal applications; commonly used leasing terminology; types of leases and lease structures; historical information; and residual forecasting.

Equipment leasing is a creative way to obtain the use of an asset without making a sizeable capital outlay to purchase it. In the business world, owning property is not usually an objective; rather, the goal is to obtain the rights to use an asset to generate economic benefits. The ability to use an asset to produce economic benefits is a fundamental element of free enterprise.

All leases use either tax or accounting rules or a combination of both. The lessee's objectives of taking advantage of tax depreciation, removing assets from the balance sheet, and obtaining the lowest implicit financing rate usually drive the selection of a lease structure.

The evolution and growth of leasing is connected to the growth that has taken place in free economies. It is interesting to note that leasing is currently becoming a fundamental part of the economies of developing and former socialist countries.

Definition and Terms

Leasing versus Renting

The term *leasing* is often used interchangeably with the term *renting*. There is little difference between the two terms other than the duration of the contract. A lease tends to have a longer term than a rental. The rights of the lessee tend to be better detailed and documented than the rights of a renter. A lease typically details various remedies and is generally noncancellable for the duration of a specified term.

For example, consider a five-year lease that includes a cancellation penalty clause stating that a breakage fee of 10 percent of the original asset value being leased would be assessed along with all the remaining uncollected rents. Under a rental agreement, however, normally no penalty clause is associated with the contract.

Lessee and Lessor

The *lessee* is the individual or company that is using, or has the rights to use, an asset for a specified period of time in accordance with the terms and conditions of the lease. The *lessor* is normally the legal owner who pays for the asset and permits the lessee to use the asset in accordance with the lease contract.

Residual Value

Another term that is common to leasing is *residual value,* which is the value remaining after part of the asset's life has been consumed. Depending on the particular lease agreement, the term *residual value* can have slightly different meanings, which are discussed more completely in the residual forecasting section of this chapter.

History of Leasing

Leasing has been in existence, in one form or another, since the beginning of civilization. Though the structures have changed over time, the basic concept of leasing remains the same: the right to use an asset without purchasing it.

The earliest forms of leasing were relatively straightforward and centered on ships or vessels used to enhance trade. These early leasing structures were based on the premise that a portion of cargo would be given to the owner of the vessel in return for its use. The owner would then sell that portion of the cargo to obtain a return on his investment. This continued to be the main form of leasing until about 200 B.C., when the first ship charters were introduced. These charters permitted various governments to obtain private ownership of ships for use in private commercial enterprise. Ship charters continued for centuries as the dominant form of leasing. The only major changes that took place were that equipment types grew to include wagons, horses, and other transportation-related equipment.

From the mid-1700s through the 1850s, the United States enacted laws relating to bailments for hire, which is an archaic term for leasing personal property. These bailments were the forerunners of modern-day leasing.

In the 1860s, the "New York Plan" was developed for financing railway rolling stock and was the predecessor to equipment trust certificates. Trust certificates are documents that show evidential ownership in the equity of property. The New York Plan was followed by a more sophisticated version of trust certificate called the Philadelphia Plan. At this time a tremendous change—industrialization—was occurring in the United States. The demand for equipment was growing to such a point that new means of financing were needed. To relieve the pressure on capital markets, laws and regulations were introduced to foster new ways to raise capital and meet the demands of the expanding economic base.

In the 1880s, the Singer Sewing Machine Company wanted to help customers purchase sewing machines. They introduced the first time-payment program, which is similar to today's conditional sales agreement: customers could purchase a sewing machine by paying a small deposit and making weekly payments. When the machine was paid off, the title transferred to the customer.

Short-term leases for railcars were introduced in 1908, creating an alternative to the traditional full payout leases. Full payout leases are leases whereby the full cost of the asset is recovered during the lease term. Short-term leases rely on renewals of the lease term or subsequent leases to another lessee to recover the cost and make the lessor whole. Short-term leases were the predecessors of operating leases, which came about 20 to 30 years later. Eventually operating leases expanded to include other types of assets.

In the 1920s and 1930s, machinery manufacturers dominated the leasing industry in an attempt to control the used equipment market. Then in the 1940s, the United States created the Lend-Lease Act to supply the World War II allies with needed war equipment; the Act allowed defense contractors to write off the cost of the equipment purchases over the length of the defense contract. The Act strengthened the role of leasing in the wartime economy and proved to be a foundation for future laws dealing with leasing.

In 1947 the Supreme Court established the cornerstone of leveraged leasing by ruling in *Crane v. Commissioner* that an owner could include in his tax basis amounts borrowed on a nonrecourse basis[1] that were secured by the value of the property.

During the remainder of the 1940s and early 1950s, several rules defining the treatment of leases were clarified and expanded. Accelerated depreciation was introduced, and in the mid-1950s, the Revenue Ruling 55-540 of the IRS provided rules differentiating a lease from an installment sale. This ruling facilitated growth in air transportation and the car rental industry.[2]

In 1955, the enactment of the National Banking Act permitted banks to broaden their scope of business. The year 1956 marked the passage of the Bank Holding Company Act, which permitted banks to establish subsidiaries and allowed them to offer a wider range of services.

The Internal Revenue Act of 1962 liberalized depreciation guideline lives and created a 7 percent Investment Tax Credit (ITC), which increased the benefits of asset ownership. In 1963, the Controller of the Currency further stimulated leasing by issuing Interpretive Ruling 7.3400 (IRS IR 7.3400), which allowed banks to engage in the leasing business.

Over the next several years, various IRS rulings affected leasing, rulings that further defined depreciation, repealed the ITC, and then reinstated it. Leasing flourished when the ITC was introduced and declined when it was temporarily rescinded. During this period financial institutions were asking for more specific guidelines concerning their leasing activity. In 1974, the Federal Reserve Board issued Regulation to provide guidelines for bank holding companies and their nonbanking subsidiaries engaged in personal property leasing.

In 1975 the ITC was raised from 7 to 10 percent; and the IRS issued Revenue Procedures 75-21 and 75-29, which clarified tax rules for leveraged lease transactions that constituted true leases. These rulings, coupled with Revenue Ruling 76-30, formed the foundation of current-day leasing. Revenue Ruling 76-30 restricted leases of special use property and further refined certain at-risk rules for leasing by individuals. These rulings are in place today and govern the way asset leasing is performed in the United States. Subsequent IRS rulings and Financial Standards Board statements have further clarified and expanded lease structures.

Present-Day Leasing

In 1998 the Equipment Leasing Association of America (ELA) reported that over $200 billion worth of equipment was leased and that leasing had become the most common form of financing used by industry for modernization and expansion.[3] The ELA also reported that the leasing industry is continuing to grow at an annual rate of about 10 percent. This rate of growth was maintained even during the late 1980s following the permanent repeal of the ITC in 1987 and during the U.S. recession of the early 1990s. When the ITC was eliminated, many in the leasing industry initially believed that leasing would be substantially affected and might even disappear. Leasing did not disappear, although a great deal of consolidation occurred, resulting in stronger, more established leasing companies. As the economy recovered from the recession, the leasing industry grew even faster. The current consensus is that leasing will continue to grow to meet the increasing demand of companies for capital equipment.

Lease Structures

Leases are usually based upon one of the following four lease structures:

- true lease,
- conditional sales agreement,
- TRAC lease, and
- leveraged lease.

These all have one fundamental thing in common: they are contracts obligating the user of the asset—the lessee—to pay the legal owner—the lessor—for the right to use the asset during the lease term. These lease structures differ in the manner in which the rights of ownership are allocated between the parties.

True Lease

A *true lease* is the most commonly used type of lease. It is also called an *operating lease, conforming lease, tax lease, genuine lease,* or just plain *lease*. A true lease is one in which the lessor is the legal owner holding title to the asset being leased; as such, the lessor has the rights and benefits of ownership for tax purposes. The lessee, on the other hand, has the rights to the asset that the lessor conveys through the lease. Interpretations of various IRS rulings suggest that a lease transaction should meet the following criteria to qualify as a true lease:

1. At the beginning of the lease term, the leased asset must have a projected fair market value, at the expiration of the lease term, of an amount greater than or equal to 20 percent of the value of the leased asset at the inception of the lease, excluding from consideration the effect of

inflation and/or deflation and any cost to the lessor for removal.

2. The leased asset is projected to have the longer of (1) at least 20 percent of its expected normal useful life (the life projected at the inception of the lease) remaining at the end of the base lease term; or (2) a remaining normal useful life of at least one year at the end of the base lease term.
3. The lessee cannot have a right to purchase or renew the leased asset for a price that is less than its fair market value.
4. The lessor cannot have a right to force the lessee to purchase the leased asset at a fixed price.
5. The lessor must have a minimum unconditional "at risk" investment equal to at least 20 percent of the value of the leased asset at all times during the lease term.
6. The lessee must not furnish any part of the purchase price of the leased asset, nor have loaned or guaranteed any indebtedness created in connection with the acquisition of the leased asset by the lessor.
7. The lessor must show that the lease transaction was entered into for profit, apart from any tax benefits resulting from the transaction.

If the proposed lease does not meet all of these conditions, it does not qualify as a true lease. A true lease provides benefits to both the lessor and lessee, making it an attractive form of leasing. The benefits are measured by both parties to determine the "all-in costs" of a leasing transaction. By examining the benefits and costs, the lessee is able to decide whether it is more prudent to lease or buy.

The lessee's benefits can include the following:

- lower cost of funds,
- hedge against inflation,
- little or no initial cash outlay,
- off-balance-sheet treatment,

- cash flow improvement,
- lease payments made from pretax earnings,
- level payments,
- no public disclosure,
- improvements on asset-to-earnings ratios, and
- the shift of risks associated with asset ownership, such as obsolescence, to the lessor.

The lessor's benefits include the following:

- retained tax benefits that can be sold to another party or retained to offset earnings,
- a higher rate of return commensurate with the added risk, and
- a diversified loan portfolio and, thus, a spreading of risk.

Conditional Sales Agreement

A *conditional sales agreement* is also referred to as a *lease intended as security*. It differs from a true lease in that the asset is considered the lessee's property. Essentially, this type of lease is nothing more than a time purchase contract. The lessee is the owner of record for tax purposes and possesses all rights of ownership, just as if he had made a direct purchase. The primary advantage to the lessee is that the lessee does not incur the full financial impact at the date of acquisition but does obtain the rights of ownership for a minimal investment. For tax purposes, the lessee has the right to take all allowable tax depreciation, although time payments are being made. This lease structure permits the lessee to purchase the asset at the end of the term for a nominal amount. With this type of lease, there is no discount in the interest rate offered to the lessee, because the lessor does not retain any tax benefits.

TRAC Lease

A *terminal rate adjustment clause (TRAC) lease* is structured as a single-investor lease and is restricted to over-the-road vehicles. Congress created these leases as a hybrid type of lease containing elements of both true leases and conditional sales agreements. The lessor is permitted to claim the tax benefits of ownership, and the risks of ownership are shifted to the lessee.

With a TRAC lease, the last lease payment can be adjusted upward or downward to make up for any difference between the originally estimated value of the vehicle, upon which the initial lease payments were based, and the actual value of the vehicle at the end of the lease term. If the value of the vehicle is higher than originally estimated, the lessee and lessor may share the difference, or it may be credited solely to the lessee. If the value is less than the original projected value, the lessee is required to compensate the lessor for the shortfall. TRAC leases are popular with car and truck leasing companies, such as Hertz and Avis, which lease vehicles from a lessor, such as an automobile manufacturer, then lease them to their customers.

Leveraged Lease

A *leveraged lease* is conceptually the same as a single investor lease: the lessee decides what equipment is needed and negotiates the terms and rates with the lessor in the same manner. The lessee also negotiates other terms, such as use, maintenance, renewal, return, and/or purchase options, as he would do in a single investor lease. A single investor lease, however, involves only two parties: the lessee and the lessor. In contrast, a leveraged lease includes at least three parties: a lessee, lessor, and a long-term debt holder.

Usually several lessors and several debt holders are involved in a given transaction, as well as a trustee who monitors each investor's interests and distributes each party's share of the lease payment. The lessor in a leveraged lease becomes owner of the asset by providing 20 percent to 50 percent of the necessary capital. The balance of the capital is borrowed from institutional investors on a nonrecourse basis to the lessor. This loan is secured by a first lien on the asset and the assignment of the lease and the lease rental payments.

The lessor in a leveraged lease can claim the tax benefits normally attributable to ownership even though he has provided less than 100 percent of the capital for the transaction. This is the "leverage" in a leveraged lease. This structure increases the tax benefits and debt but simultaneously increases the risk of the transaction and hence the rate.

Off-Balance-Sheet Loans

Recently another type of financing vehicle has gained popularity, the *off-balance-sheet loan* (OBSL). It is viewed as a true lease for financial accounting purposes and a conditional sales agreement or loan for tax purposes. It is treated as a loan for tax purposes with the lessee taking depreciation and carrying the asset on its books for tax purposes. For financial accounting purposes, however, the assets are only footnoted and thus are treated as leased assets.

In OBSL transactions, lessees have many of the same options that they would normally have in a true lease. The purchase price at the end of the term cannot be a bargain purchase or less than fair market value. OBSL transactions come in a variety of slightly different versions but maintain two common elements: the lessor is at apparent risk and the lessee is offered the option to purchase at the end of the term.

Appraisals for Leasing Purposes

Appraisals play a significant role in leasing. The choice of lease structure determines the function, purpose, and type of appraisal. It is important to have a basic understanding of the various lease structures. Appraisals not only support the lease structure but are frequently required at other times during the term of a lease. Therefore, it is necessary to have a basic understanding of the three phases of the lease life cycle, the structure phase, the lease period, and the end of lease.

Structure Phase

The lease life cycle begins when a potential lessee requires additional assets and needs to identify a method to finance the acquisition. This phase is referred to as the structure phase. Lessors are requested to provide proposals describing potential lease structures.

These proposals are reviewed, and one or more of the prospective lessors is asked to provide a formal proposal. Assuming that the structure and pricing meet the lessee's objectives, a proposal is accepted. This acceptance is referred to as being mandated or awarded. From this point until the closing or signing of the final lease documents, the lease language is refined, values of the assets are reviewed, and final adjustments are made.

During the structure phase, the most common need for appraisal services is to determine the expected normal useful life of the asset and to estimate residual values at the expiration of the lease term. The IRS requires an appraisal to determine the asset's residual value. As previously mentioned, leased assets must satisfy the requirements of various IRS regulations to qualify for leasing treatment for tax purposes. The lessor will often depend on an appraiser to help determine whether the assets meet the following requirements:

- The asset must not be "limited use property," that is, the asset must be able to be used by parties other than the lessee.
- At the end of the lease term, the leased asset is projected to have a remaining useful life equal to at least 20 percent of its remaining useful life at the inception of the lease term or a remaining useful life of at least one year.
- The asset must have a residual value at the end of the lease term that is equal to at least 20 percent of its value at the inception of the lease.

For sale/leaseback transactions, an appraisal is required to determine the asset's value at the time of the sale to the lessor. A projection of the residual value is also required. In a sale/leaseback transaction, the lease structure depends on the value conclusions as well as the accounting treatment of the transaction for the lessee.

For synthetic leases, such as OBSLs, an appraisal is needed to determine the asset's residual value and to project its value annually.

Lease Term

The *lease term,* also known as the *lease period,* is used to define the period from the lease commencement date to the lease expiration date. During the lease period, appraisals may be needed for a variety of reasons: credit analysis; Financial Accounting Standards Board guideline 13 (FASB-13) review; potential sale of the lease or residual position; or early lease termination. Each need requires a different type of appraisal.

Appraisals performed for credit analysis generally require the asset to be valued at orderly liquidation value, forced liquidation value, or both. The credit department of the financial institution needs to know what can be reasonably recovered from the sale of the asset; this information helps the department decide to either sell or wait for better market conditions.

FASB-13 requires that leasing companies annually review the residual risk positions of their leases. This rule states that if the future value of the asset drops below its booked residual value, and the downturn is other than temporary, the leasing company must, in the year it discovers the change, establish a reserve to offset the potential loss. This can have a significant effect on earnings if the residual risk positions cannot be supported. Appraisals performed to establish residual values should be consistent with the terms and conditions of the lease agreement and the definitions of value contained therein. The reason for this requirement is to ascertain that the financial position of the leasing company is properly reported.

A leasing company often buys or sells individual leases or lease portfolios either to manage the company's income or to adjust asset concentrations in its portfolio. These transactions require appraisals to properly analyze the value of the assets.

An asset may also be sold when its value has increased. In this case, the new lessor assumes a higher base, the selling company recognizes a gain, and the lessee receives a better rate.

The appraiser should remember that real money changes hands in a leasing transaction and that leasing is based on contracts. A leasing appraisal must be consistent with the terms, definitions, and conditions in the lease. Performing an appraisal without an understanding of the lease's structure and terms can lead to erroneous value conclusions.

End of Lease

A lease is a contract between the lessor and lessee. The lessor transfers certain rights of ownership to the lessee by means of the lease contract. The lessee exercises those rights for a prescribed period defined as the lease term. In some leases the legal owner conveys all rights of ownership to the lessee except those rights that are specifically identified in the lease as "held back."

At the end of the lease term, the rights of ownership that were temporarily conveyed through the lease to the lessee revert to the legal owner of the asset. In many cases, however, this is not simply a return of the asset to the lessor. In most true and leveraged leases, certain options are available to the lessee at lease expiration. These options generally include renewing the lease, purchasing the asset, or simply returning the asset to the lessor. The renewal and purchase options generally trigger the need for an appraisal.

Renewal Option. A renewal option permits the lessee to renew the lease for another prescribed period. The rental for the renewal period is typically based on the asset's then-current fair rental value as defined in the lease. The typical definition of fair rental value is the price that a willing and informed owner would accept for renting an asset to an informed and willing buyer in an arm's-length transaction for a prescribed period of time. The lease may outline several different renewal options, each for a different period of time and possibly with a different basis of value.

Purchase Option. Most true leases provide a purchase option. This option permits the lessee to purchase the asset at the expiration of the initial lease term. If chosen, the option must be exercised in accordance with the terms and conditions of the lease. The terms and conditions provide the definition of value used in the lease. Normally the purchase price is based on the current fair market value of the asset. Any appraisal that is done must use a value premise that is consistent with the definition in the lease. For example, the lease may define the purchase price as the fair market value of the asset installed at its present location, or it may define it as the fair market value of the asset de-installed, crated, and ready for shipment.

IRS regulations require that, at the end of a true lease term, the lessee must be able to purchase the asset at a price based on its then-current fair market value. Failure to meet this requirement could result in a ruling by the IRS that the lease agreement was not a true lease, thereby requiring that the transaction be reclassified a conditional sale, triggering tax penalties and requiring that the lease payments be recalculated.

Many leases define value in a way that differs, sometimes in important ways, from the value concepts commonly used by appraisers (including the value concepts used in this book). In appraisals for leasing purposes, the definitions of the lease take priority. Therefore, it is essential that the appraisers read and understand the definitions of value and other terms used in the lease. Failure to do so may result in an appraisal that reaches inappropriate or erroneous conclusions.

Often at the end of the lease, the appraiser is asked to provide additional information about the leased asset, including whether the asset was properly maintained. If the appraiser finds the asset's condition is lower than that required by the lease, the appraiser must determine the cost of restoring the asset to the required condition.

Residual Forecasting

Residual value projection is an important component in the structure of a lease, but it is also one of the least understood concepts. Essentially, the term *residual value* means the value remaining after the asset reaches the end of its normal useful life. It can also refer to the fair market value at a defined future point in time.

For leasing purposes, a good starting definition of residual value is the value of an asset at the end of the lease term. Different leases may further define the term. For example, one lease may define residual value as fair market value under a removal concept, that is, the value of the asset properly removed, crated, and ready for shipment. Another lease may define residual value as fair market value of the asset installed and ready to use. Yet another lease may define residual value as the fair market value of the asset properly removed, shipped, modified, and installed at another site.

Residual values are based on forecasts of future events. Like all forecasts of future events, residual values are inherently somewhat speculative. Nevertheless, by applying the proper techniques for obtaining, analyzing, and interpreting facts and data, the risk of speculation can be minimized.

Residual values are a critical component of almost every lease structure. They are especially critical to the lessor in the pricing phase of most tax-related lease transactions. The primary objective of the lessor is to receive an adequate return-on-investment to compensate for the cost of funds and the inherent risk associated with the lease transaction. In many cases the stream of payments over the lease term will pay only for the amortization of the asset value and cost of funds, and the residual value represents payment for the risk. At lease expiration, the lessor hopes to receive a reasonable value for the asset to earn a reasonable return-on-investment.

Residual values are required to determine compliance with IRS Ruling 55-540 and Revenue Procedure 75-21, which state that the asset (1) must have at least 20 percent of its value (as of the inception of the lease) remaining at the end of the lease term, and (2) cannot be leased beyond 80 percent of its normal useful life. Proper residual forecasting mitigates future losses from overaggressive residual value assumptions.

Factors Affecting Residual Value

Several factors affect residual value. These factors fall into three basic categories: economic, asset-specific, lease-specific.

Each category should be analyzed alone and in relation to the others.

Economic Factors. The performance of the general economy affects residual value. If the economy is in a recession, the marketability of used assets is usually negatively affected. Projections made during a recession could significantly underestimate future values, because most secondary market data would be influenced by the existing recession. The opposite could happen if projections are made during a boom economy. As history demonstrates, economies tend to run in cycles, which can be global, regional, national, or industry-specific. Failure to understand or compensate these dynamics can result in the forecasting of inappropriate residual values.

Asset-Specific Factors. Asset-specific factors are those directly connected to the asset itself. These include the manufacturer's reputation, the popularity of the manufacturer as measured by its market share, and the equipment brand or model. The manufacturer's reputation has a significant influence on the marketability of the asset in the used equipment market. If a manufacturer has a reputation for building inferior products or does not maintain an adequate stock of replacement parts, the marketability of that manufacturer's products can be negatively affected. The manufacturer's market share indicates the acceptability of its products and the desirability of ownership. If the manufacturer's product represents only a small segment of the market, the asset may have lower future values. Production levels also influence residual values because the secondary market is affected by supply and demand. Access to and location of secondary markets is another important factor in residual valuation.

In most leases, the asset funding typically includes both direct, "hard" costs and indirect, "soft" costs. In leasing, direct costs are generally the actual equipment costs, add-ons, and any tangible features. Indirect costs contribute value if the asset is installed, but they do not exist if the asset is removed. Examples of indirect costs are freight, tax, installation, operating licenses, and related items.

Each industry has its own product cycle. Some industries undergo radical changes in technology over a five-year period, whereas others change little over 20 years. It is important for the appraiser to understand where a particular industry was in a given cycle when the lease originated and where in the cycle it may be when the lease terminates. For example, in the 1960s computers had developmental cycles of about ten years. Cycles continued to become shorter until in the 1980s they reached about one year. Eventually manufacturers realized they could not recover their research and development costs because consumers were unwilling to buy new systems each year, with the result that the computer market now operates on a three- to five-year cycle.

Other factors must be addressed when projecting residual value; failure to do so could result in erroneous conclusions. Some of these factors are

- Is an operating license required to operate the asset and is it transferable?
- Are environmental issues or regulations pending that may affect future marketability?
- What is the environment in which the asset will be used?

Lease-Specific Factors. Lease-specific factors are contractual in nature and inherent in the lease itself. The sections of the lease providing lease-specific information are those relating to use and maintenance; purchase and renewal; and returns.

The use and maintenance section of the lease details how the asset must be maintained and any restrictions imposed on its use. In many cases, the lease requires the lessee to maintain the asset in accordance with the original manufacturer's recommended standards and procedures. Other leases may state that the asset must be maintained to "prudent" industry standards or may limit the annual number of hours the asset can be used.

The purchase and renewal section specifies

- how the purchase price is determined;
- the definition and premise of value;

- when the lessee must notify the lessor of his intent to renew, purchase, or return the equipment; and
- the details governing any storage period that may be necessary, which is usually provided at the lessee's cost.

In addition, the return provision provides important directions as to how the asset should be valued. For example, it may answer such questions as whether the lessee or lessor will bear the cost of removal, whether any performance guaranties are included, or whether the asset has to meet certain performance specifications before it is returned. These lease-specific factors all affect the future value of the asset.

Projecting Residual Value

There is no one correct method for projecting residual values. Currently several different methods are used, including those based on historical transaction data and those based on the anticipated income stream the assets are projected to produce.

Most residual forecasting is based on historical transaction data. Past performance is a good guide to future performance as long as the appraiser understands the factors that influence performance. Care must be exercised when using historical data; the data and its relationship to the current set of circumstances must be understood.

Historical transaction data may be analyzed in two ways. The first is to compare the historical data to current prices and develop ratios on the relationships. This method assumes that the current economic conditions and other factors affecting value will remain the same in the future. The second method adjusts current market information by removing the effects of inflation over time and then comparing the used costs to the historical costs to establish ratios. This method results in a constant dollar comparison of the cost new and the cost used. By developing a series or a matrix of such relationships over time, anomalies can be identified and researched further. The matrix can provide additional information, such as

- asset life cycles,
- when new models are introduced and their effect on preceding models, and
- when in the life cycle the asset experiences its greatest loss in value according to a graph of the asset's declining value over time (decay curve).

Residual forecasting is not an exact science. As with all quantitative problems, supportable conclusions can be reached if research is conducted and if relevant data are collected and analyzed using proper techniques. Care must be taken in gathering and understanding the data. Data that are not understood are dangerous and can lead to inaccurate projections.

Key Points

- Equipment leasing is a creative way to obtain the use of an asset without making what can be a sizeable capital outlay to purchase it. In the business world, owning property is not usually an objective; rather, the goal is to obtain the rights to use an asset to generate economic benefits. The ability to use an asset to produce economic benefits is a fundamental element of free enterprise.

- All leases use either tax or accounting rules or a combination of both. The lessee's objectives of taking advantage of tax depreciation, removing assets from the balance sheet, and obtaining the lowest implicit financing rate usually drive the selection of a lease structure.

- The lessee is the individual or company that is using, or has the rights to use, an asset for a specified period of time in accordance with the terms and conditions of the lease. The lessor is normally the legal owner who pays for the asset and permits the lessee to use the asset in accordance with the lease contract.

- Leases are usually based upon one of the following four lease structures: true lease, conditional sales agreement, TRAC lease, and leveraged lease. These all have one fundamental thing in common: they are contracts obligating the user of the asset, or lessee, to pay the legal owner, or lessor, for the right to use the asset during the lease term. These lease structures differ in the manner in which the rights of ownership are allocated between the parties.

- Many leases define value and other terms in ways that differ, sometimes in important respects, from the value concepts commonly used by appraisers (including the value concepts used in this book). In appraisals for leasing purposes, the definitions of the lease take priority over other definitions. Therefore, the appraiser must read and understand the definitions of value and other terms that are used in the lease. Failure to do so may result in an appraisal that reaches inappropriate or erroneous conclusions.

- For leasing purposes, a good starting definition of residual value is the value of an asset at the end of the lease term. Different leases, however, often further define the term in ways that are essential for the appraiser to understand.

- Residual values are based on forecasts of future events. Like all forecasts of future events, forecasting residual value is inherently somewhat speculative. Nevertheless, by applying the proper techniques for obtaining, analyzing, and interpreting facts and data, the risk of speculation can be minimized.

- Several factors affect residual values. These can be grouped into three categories: economic, asset-specific, and lease-specific.

- There is no one correct method for projecting residual values. Currently several different methods are in use, but historical transaction data are the foundation for most residual forecasting.

- Residual forecasting is not an exact science. As with all quantitative problems, supportable conclusions can be reached if research is conducted and relevant data are collected and analyzed using proper techniques. Care must be taken in gathering and understanding the data; data that are not understood are dangerous and can lead to inaccurate projections.

Additional Reading

Amembal, Sudhir P., and Terry A. Isom. *The Handbook of Leasing; Techniques & Analysis.* New York: Petrocelli Books, 1982.

BankAmeriLease Companies. *The Complete Guide to Leasing.* San Francisco, Calif.: Bank of America, NT&SA, 1987.

BankAmeriLease Companies. *The Red Book of Leasing.* San Francisco, Calif.: Bank of America, NT&SA, 1985.

Contino, Richard M. *Legal and Financial Aspects of Equipment Leasing Transactions.* Englewood Cliffs, N.J.: Prentice Hall, 1979.

Notes

[1] Nonrecourse debt is debt that is secured solely by the asset being leased; there is no secondary source of repayment.

[2] Revenue Ruling 55-540; 1955-2 CB39.

[3] *Equipment Leasing Today* (January 1995). Arlington, Va., American Association of Equipment Lessors.

9

Use of Personal Computers in MTS Valuation

Objectives:

1. Define the components of personal computer (PC) hardware.
2. Describe the standard desktop PC configuration.
3. Suggest components for purchasing a computer system.
4. Define types of software systems and application software.
5. Outline online communications.

As in other professions, the appraisal profession has undergone many changes in the past few years. Not long ago the MTS appraiser's work was predominantly performed for insurance purposes. Today, the scope of work has expanded. The appraiser now prepares appraisals for a number of different reasons, including financing, leases, sales, mergers, and dissolution. Because of this expanded scope, more sophisticated valuation tools are needed.

Less than a decade ago, the MTS appraiser had weeks, if not months, to prepare an appraisal. Today, the period to complete an appraisal can be days or mere hours. The information age has created new client needs and expectations that require the appraiser to work with more speed and efficiency.

Because of the need for more sophisticated and timely valuations, the use of PCs in MTS valuation has become an everyday occurrence. The advent of PCs has marked the beginning of a new era in the way appraisers collect, analyze, prepare, and report information. Few appraisal assignments are performed that do not require some use of PCs in the valuation phase, reporting phase, or both. The appraiser, in an effort to increase productivity, now uses the PC more frequently than all other office equipment, supplanting such mainstays as the typewriter and calculator.

This chapter provides an overview of the basic ways PCs are used in the MTS valuation process. General information about computer hardware and software is discussed and examined in the context of MTS valuation. Two things should be noted at the outset: (1) Although this chapter refers specifically to IBM-compatible computers and software, because they are most commonly used by appraisers and their clients, nearly all of the information discussed herein is also applicable to Apple Macintosh systems; (2) this chapter was written in early 1999 and, obviously, references to "current" or "today's" speed, capacity, and so forth are based on the standards then existing.

PC Hardware

Desktop Computers

Personal computer hardware refers to the equipment components that make up a PC system. This section will focus on the hardware associated with PCs, as opposed to larger computers such as mid- and mainframe systems. Major components of the PC discussed include the following:

- central processor,
- hard drive,
- floppy drive,

- CD-ROM drive,
- memory,
- monitor and video adapter card,
- keyboard,
- mouse, and
- modem.

Computer hardware is always evolving, with technological breakthroughs occurring on a frequent basis. In computing, what is state-of-the-art today will be obsolete tomorrow. Although the following components will likely continue to exist in PC systems for the foreseeable future, their speed, size, and efficiency will surely change.

Central Processing Unit (CPU). The CPU is the device or microprocessing chip that serves as the computer's brain. It interprets and processes all the information necessary for the computer to operate. The Intel chip is the most common.

Since the early 1980s, Intel (the leading chip manufacturer as of this writing) has developed several generations of CPUs for PCs. The first was the 8088 CPU that had the equivalent processing power of roughly 30,000 transistors. The next generation of chip, the Intel 80286, was developed in the mid-1980s for use by IBM to power its AT class of machines. In the late 1980s, the 80286 was gradually phased out in favor of the 386 CPU. This chip had the power of 275,000 transistors.

By the early 1990s, Intel had developed the 486 CPU. This processor has since given way to the current generation of CPU known as the Pentium chips. Introduced by Intel in 1994, Pentium chips are now the standard in new PCs. Cirrus Corporation, endorsed by IBM, and the AMD Corporation have developed 586 and 686 processors to compete with Intel's Pentium chips, and there may be others. These new chips have processing speeds rated in terms of megahertz. A megahertz (MHz) is a measure of frequency equivalent to one million cycles per second. The Pentium and 686 chips now operate (in early 1999) in excess of 400 MHz. This is a dramatic increase from the 8088 chip of the early 1980s, which operated at 4 MHz. The MMX feature has been added to many recent Pentium chips or motherboards. It simply enhances multimedia features, making sound and recording features better.

Hard Disk Drive (HDD). The *hard disk* or *hard drive* is the device that stores and records data or information in the form of a file. It functions in a manner similar to that of the record player component of a stereo system, but, unlike the record player, it can also record information. The first PC with a hard disk was the IBM XT, which, when introduced in 1983, had a storage capacity of ten megabytes. (A megabyte or "MB" is the equivalent of 1,048,576 bytes or characters of information.)

Today a 10 MB hard disk is a relic. Most new PCs come with hard disks measured in gigabytes. (1,000 MB is equivalent to a gigabyte.) Storage capacity is a significant factor in the cost and desirability of a hard disk, but its performance is also a factor.

Performance is measured by the speed at which a hard disk accesses data. Because software programs frequently read and write to the hard disk, the performance of the hard disk can have a major impact on the overall performance of a PC. Measured in terms of milliseconds (ms), the performance of a hard disk is determined by its average access time, the time it takes for the read-write head to move from one random spot on the disk to another. When first introduced in the IBM XT, hard disks had an access time as high as 65 ms. Today, an access time of 15 ms is considered slow.

Traditionally, hard disks have been built into the computer system. Recently, the popularity of the removable hard disk has increased dramatically. These units currently can store anywhere from 100 MB to more than a gigabyte. Because they can be removed, the removable hard disk cartridge is the ideal solution for fast, portable, unlimited storage. Some users have now elected to use a removable hard disk as their primary drive—a fast, simple way to attain increased storage and security. New, removable hard disks have access times equal to previous internal units, and data transfer rates are extremely fast.

Floppy Disk Drive (FDD). The *floppy drive* is the device that reads and writes information to a removable diskette, a flat piece of Mylar that magnetically captures information. Typically the floppy drive is used to install software to the hard disk and transfer files between PCs. In early PCs, floppy drives were used to run all the software needed to make a machine operate. They used 5.25-inch floppy diskettes that were able to hold 360 kilobytes of data, less than one-half of a megabyte. In the mid-1980s, floppy drives began to use high-density diskettes with a capacity of 1.2 MB. Today, the standard floppy drive uses a 3.5-inch diskette and has 1.44 MB of capacity. Floppy drives are commonly used to exchange data files from one user to another and to back up information stored on the system's hard drive.

CD-ROM Drive. Like the hard drive, this component allows access to stored data. A compact disc (CD), the same type used to store music, is inserted into a CD-ROM (read-only memory) drive and then "played" to the PC. A single CD can hold up to 650 MB of information. Unlike a hard disk, a CD is easily removed or changed. Therefore, with a CD-ROM drive, a PC can become a desktop library. In addition to storing data, such as words and numbers, CDs can also store sounds, graphics, photographs, animation, and videos. (To play sounds from the CD-ROM drive, a sound card and speakers are required.) Because CD-ROMs hold such large amounts of data, they are excellent for storing publications. A single CD can replace entire series of books—encyclopedias, reference manuals, telephone directories, and road atlases, among others.

Until recently, CD drives had "read only memory" (ROM), where the data were permanently encoded and could not be erased or modified. In late 1995, Hewlett Packard Corporation introduced the first reasonably priced CD drive that could modify, replace, and record data as well. This development will change the way information is commonly stored and retrieved. There is one drawback with the CD drive: its performance. Compared to a hard disk, a CD drive is relatively slow. Speed notwithstanding, the CD drive and advances to it will continue to enhance the multimedia experience.

Now on the market are digital videodisks (DVDs), similar in appearance to CDs but with far greater storage capacity. A single DVD is capable of storing an entire feature-length movie at near lifelike resolutions.

Random Access Memory (RAM). This term refers to chips that dynamically store information that a computer can access at any time while running an application (as distinct from information stored on a disk). Memory is measured in bytes. In PCs, RAM memory resides in two distinct areas or addressable locations of the chips. They are usually called base memory and HMA (high memory areas). Base memory is the first megabyte of RAM, which is divided into the conventional memory and UMB (upper memory blocks). Conventional memory is generally the first 640 kilobytes of base memory. All current action must run from conventional memory. The UMB is generally reserved for system and hardware instructions; however, most current operating systems make unused areas of UMB available as additional conventional memory. The remainder of RAM is usually called extended memory.

The more RAM a PC has, the better it is able to process large amounts of information. As of this writing, PC manufacturers install a minimum of 64 megabytes of RAM in their systems, with 128 or more megabytes quickly becoming the standard.

Monitors and Video Adapter Cards. These components are hardware components that have also experienced technological advancement. Monitors are the devices on which images are generated. Like televisions, they are measured in terms of diagonal screen size (in inches).

Video adapter cards are circuit boards that fit into a PC and generate the signals needed for monitors to display images. In the early 1980s, only monochrome monitors and adapters were available. Since then, there have been color graphic monitors and adapters (CGA), enhanced graphic adapters (EGA), video graphic adapters (VGA), and, currently, super video graphic adapters (SVGA). Super video graphic monitors and adapters provide crisp and clear text and images and support graphic resolutions as high as 1024 x 768 pixels and up to 256 colors. A pixel refers to the number of vertical and horizontal lines that can be viewed. Dot pitch, the distance between dots on the screen, has also evolved in monitors. The smaller this distance, the sharper and crisper the image displayed. A dot pitch of 0.28 mm is the current industry norm for an SVGA monitor.

Keyboard. The *keyboard* is the device from which most information is entered into a computer. Numbers, letters, punctuation marks, and other characters are positioned on the keyboard in a layout resembling a typewriter. The computer keyboard differs from the typewriter in several respects. It contains special keys including function keys (F keys), an alternate key (Alt key), and a control key (Ctrl key). These keys provide access to unique functions or abilities specified by the software application being used. There are also additional keys—including page up, page down, home, and end—which allow the user to navigate to different areas displayed on the monitor. In addition, enhanced keyboards include a numeric keypad where numbers and arithmetic symbols are laid out in a manner similar to a calculator. New contoured keyboards are currently on the market, but usually an enhanced keyboard with 104 total keys is adequate.

Mouse. The *mouse* is a pointing device that gets its name because of its resemblance to a real mouse. Used with software programs that have a graphical user interface (GUI), a mouse activates commands and performs other tasks such as highlighting text, drawing figures, and moving around the monitor screen. The mouse is rolled on a pad or flat surface. The movements are translated into corresponding movements on the monitor. By pointing to an icon (a small graphical picture on the screen) and clicking one of the mouse's buttons, the user sends command instructions to the PC. Although many of the tasks that can be performed by a mouse can also be performed with a keyboard, the mouse is often faster and easier to use.

There are several other types of pointing devices, such as a wireless mouse, which uses infrared signals; the trackball, which is merely an inverted mouse; and the fingerpad, which tracks finger movements made on a small electronic pad.

Most current software programs, especially those that are Microsoft Windows–compatible, use a mouse to activate pull-down menus and perform other functions. All PC manufacturers include a mouse or some other pointing device with their systems.

Modem (modulator/demodulator). The *modem* is a device that allows computers to communicate via telephone lines. The name derives from the ability to modulate and demodulate transmitted information. With a modem, a computer is able to log on or connect to any number of online services, Internet sites, or other computers. Once connected, modems allow users to send and receive files, graphic images, audio tracks, and other types of data.

Modems can be either internal or external to a PC. An internal modem is a card or board that fits into a slot inside a PC. External modems are stand-alone components ranging in size from a cigarette pack to a small book.

Like other PC components, modems have undergone significant recent development and are smaller, faster, and cheaper than ever. Modems are rated in terms of their transfer speed or the rate at which they send and receive information. This is measured in bits per second (BPS). Currently most modems compress and decompress data during transmission using an international compression standard known as V.42. In the early 1980s, modems operating at 300 BPS were considered fast. In early 1999, a 56,000 (56 K) BPS modem, which operates 192 times faster, was the standard.

Suggested Configuration. The one certainty when it comes to PC hardware is that it will continue to evolve. The way PC technology has progressed, it is safe to assume that a system purchased today will be obsolete within five years. For practical purposes, the appraiser may consider purchasing a new system, or at least upgrading an existing system, once every three years. This will ensure compatibility with contemporary software and the ability to exchange data and files with colleagues and other business professionals.

A desktop PC configuration well suited for the MTS appraiser, as of early 1999, will likely consist of the following components:

- IBM-compatible PC with a tower or mini-tower case (tower cases are vertical enclosures that usually have room for additional peripheral devices);
- Pentium MMX processor operating at 400 MHz or faster;
- ten gigabyte or larger hard disk operating at 11 ms or faster (lower number);
- 3.5-inch high-density floppy drive;
- memory with a *minimum* of 64 megabytes of RAM;
- 32x speed CD-ROM drive;
- SoundBlaster-compatible, 16-bit sound card;
- 17-inch, or larger, super-VGA monitor with 0.28 or lower pitch;
- video adapter card with 256 SVGM color output;
- enhanced keyboard with 104 keys;
- Logitech- or Microsoft-compatible mouse; and
- internal 56k BPS modem.

Purchasing a System. Personal computer systems are available from many large retailers including Costco, Sam's Club, Wal-Mart, Sears, Office Depot, Office Max, Radio Shack, Circuit City, and the Good Guys, to name a few. Others are sold directly by PC manufacturers. These include IBM, Compaq, Micron (Zeos), Gateway, and Dell, among others. Systems can also be purchased from computer retailers including CompUSA, Micro Warehouse, CDW Computer Centers, PC Warehouse, and Frys Electronics. Finally, many local "clone" makers will custom-build a system to a buyer's specification.

The decision about where to buy should be evaluated based on price, convenience, and support. Clone makers have the best prices, but they might not be in business later to provide support or service. Large retailers have good prices, but they sell a variety of merchandise and their technical knowledge and support may be minimal. Computer manufacturers typically provide only telephone support, which can be time-consuming and frustrating. Also, items in need of repair normally must be returned to the manufacturer's service center and, depending on the component, their absence will render the system useless until returned. Computer retailers have higher prices, but their ability to support and service the equipment they sell is unmatched.

Expect to pay several thousand dollars for a packaged desktop system (excluding a printer). A packaged system is one that comes assembled and ready to run. It is better and cheaper to purchase a packaged system instead of separate components, because there is the assurance that the hardware components will be compatible.

Notebook Computers

For those who travel extensively and require portability, a notebook computer is the solution. Notebooks are small PCs that range from the size of a handheld calculator to that of an encyclopedia volume. Notebooks include built-in keyboards and monitors and can be configured to the same specifications as most desktops. Typically they use the same peripherals, including printers, external monitors, and modems.

Notebooks have now evolved to a point where they are powerful enough to serve as the main system of many business users. Using docking stations (base units into which the notebook plugs), users can have instantaneous access to all peripherals. Some docking stations include additional hard drives or backup storage devices.

Suggested Configuration. A well-configured notebook for the MTS appraiser, in early 1999, would include the following components:

- 266 MHz processor;
- 32 megabytes of RAM;
- 16x CD-ROM drive;
- 56k modem;
- 1.44k floppy disk drive;
- three-gigabyte or larger removable hard drive;

- built-in speakers; and
- 11-inch active color matrix screen.

Major notebook manufacturers currently include Compaq, IBM, Toshiba, Texas Instruments, and Hitachi. For Apple users, the Macintosh PowerBook is a compatible unit. Notebooks can be purchased almost everywhere desktop systems are sold.

One of the main drawbacks of notebooks is their high cost. Notebooks normally cost about $1,000 more than a comparably configured desktop system, depending on whether it has an active matrix or dual scan screen. An active matrix screen is a more expensive display and has the clarity and resolution of a high-quality desktop monitor. A dual scan screen is less expensive but cannot be easily viewed from an angle. Notebook computers are also relatively fragile and expensive to repair.

Software

A computer without software is like a car without gas. Software refers to the instructional information needed to make a computer operate. Software consists of programs that, when read by the hard disk, floppy drive, or CD-ROM drive, signal and direct the computer to perform meaningful tasks. In general, there are two primary types of software: system software and application software.

System Software

Before it can operate, a PC requires some type of system software (also known as an operating system). This software is responsible for controlling the fundamental elements of a PC's hardware resources, including the memory, processing unit, hard drive, floppy drive, monitor, and other peripherals. The operating system dictates how information is interpreted and is the foundation upon which application programs are built. Today, most PCs use Microsoft Corporation's disk operating system (MS-DOS). Although some PCs use alternative systems such as IBM's PC DOS, OS/2, and UNIX, Microsoft is widely recognized as the industry leader.

In addition to the operating system, PCs use a graphical user interface (GUI), such as Microsoft Windows (figure 9.1). A GUI makes a PC easier to operate. Graphical images or icons that are accessed using a mouse replace complex keystroke commands. For some time, Microsoft Corporation has been selling a network version of the Windows GUI, known as WindowsNT.

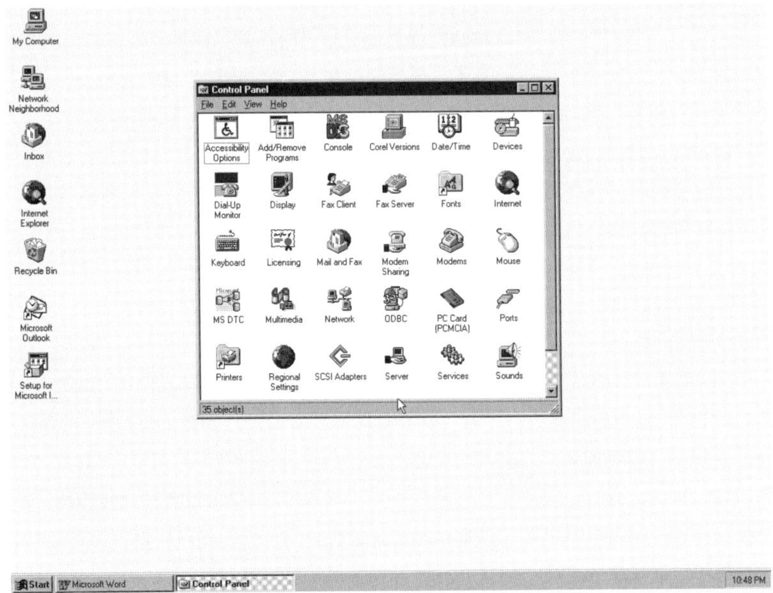

Figure 9.1. Windows 95 screen.

When purchasing system software, compatibility with any existing computer hardware is essential. Most new systems come with system software already installed. When upgrading to a new program, check with the manufacturer to see if a discounted price is available for upgrades. System software can be purchased from most of the same sources as computer hardware, directly from the manufacturer, or from a number of mail-order companies that specialize in software. Some of the larger mail-order companies include Egghead Software, PC Connection, PC Zone, Micro Warehouse, Midwest Computer Works, and Tiger Software. Many now offer online purchasing through the Internet.

Application Software

Application software refers to a computer program that allows the PC to perform a certain type of meaningful work. This work can include the manipulation of text, numbers, graphics, or a combination of these elements. The two basic versions of application software are DOS-based and Windows-based. DOS-based versions run under a PC's system software and can generate only text or character output on the computer's monitor. Windows-based applications run under a Windows GUI to allow graphical images and pictures to be displayed in addition to text. DOS applications can usually run under a Windows GUI, but Window programs cannot run under DOS alone. The major applications to be covered in this section include

- word-processing software,
- spreadsheet software, and
- database software.

Word Processor. The *word processor* was one of the first application programs developed for the PC. The word processor began in the early 1980s as an electronic typewriter in which typed information could be electronically stored. Today, many word processing programs have evolved to a point where the finished products rival those that could be produced only by a printer or publisher less than a decade ago.

Essentially, a word processor allows the user to perform all the tasks necessary to prepare a variety of documents, from simple letters to entire books. (In fact, this book has been prepared using a desktop publishing program.) The basic functions include entering, editing, deleting, copying, moving, searching, and replacing words and text. In addition, word processors can perform formatting tasks including highlighting, underlining, bolding, italicizing, and more. Most word-processing programs include a spell-check feature that scans documents and alerts the user to potentially misspelled words. Grammar checkers are also common in the major programs.

Word processors that operate under a Windows GUI use WYSIWYG technology ("what you see is what you get"). Under

WYSIWYG, the printed output is identical to what shows on the monitor screen. With a GUI word processor, typefaces and fonts (typeface subsets) are easily manipulated. In addition, pictures, charts, graphs, and other images can be inserted into documents. Advanced features include the ability to automate repetitive tasks or commands (known as macros), create multicolumn documents, generate indexes and tables of contents, draw figures and diagrams, and copy and insert text from other programs.

Word-processing software can assist the appraiser in many ways. Some of the more common uses include the preparation of letters (see figure 9.2), proposals, reports, exhibits, qualifications, client lists, questionnaires, invoices, brochures, fax coversheets, newsletters, announcements, labels, and envelopes.

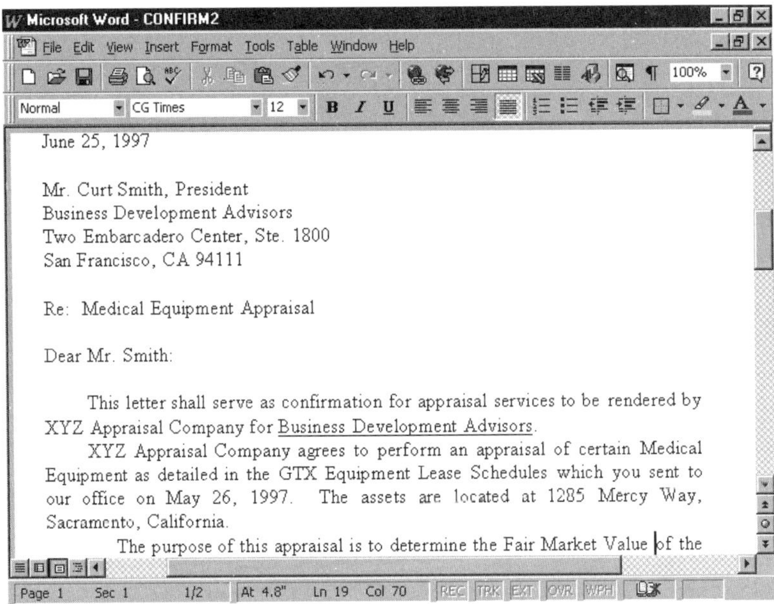

June 25, 1997

Mr. Curt Smith, President
Business Development Advisors
Two Embarcadero Center, Ste. 1800
San Francisco, CA 94111

Re: Medical Equipment Appraisal

Dear Mr. Smith:

This letter shall serve as confirmation for appraisal services to be rendered by XYZ Appraisal Company for Business Development Advisors.

XYZ Appraisal Company agrees to perform an appraisal of certain Medical Equipment as detailed in the GTX Equipment Lease Schedules which you sent to our office on May 26, 1997. The assets are located at 1285 Mercy Way, Sacramento, California.

The purpose of this appraisal is to determine the Fair Market Value of the

Figure 9.2. Sample letter — Word for Windows 97.

The appraiser can also create templates or boilerplates that streamline the preparation of the documents mentioned above. For example, a template can be made for appraisal reports that would include all of the standard or generic information common to all reports. This information might include the body of the transmittal letter, definitions of value, appraisal methodologies, limiting conditions, certifications, and qualifications. Furthermore, macros or automated keystrokes can be developed to prompt the appraiser to areas where job-specific information is required. In an appraisal report template, these areas could include the client's name and address, the effective date of the appraisal, the identification of the property, and the purpose of the appraisal. When using a word processor, the philosophy is "If you have typed it once, there is no need to type it again."

Many of the more popular word-processing programs already contain a variety of ready-made templates that can be useful to the appraiser. These include standard business letters, fax cover sheets, invoices, memorandum forms, calendars, and time reports. Depending on the appraiser's preference, it may be sufficient to modify the existing templates rather than create new ones.

Popular word-processing programs for the PC, as of this writing, include WordPerfect for DOS and Windows, Microsoft Word (DOS and Windows), Word Pro (formerly Ami Pro - Windows only), and XyWrite (DOS only). The top full-featured desktop publishers include Quark Xpress, Adobe PageMaker, and Corel Ventura, all of which are Windows-based.

Prices vary widely and currently range from under $100 for basic word processors to well over $500 for sophisticated desktop publishers. Windows 2000 and WindowsNT come with a built-in word processor called WordPad that may be enough to satisfy most appraisers' needs. If upgrading to a new word processor, perhaps from a DOS-based program to a Windows-based program, check with the manufacturer or vendor to see if "competitive upgrades" are available. Often software manufacturers offer special pricing to users of competing software to entice them to switch loyalties. Word-processing software is available from the same sources as system software.

A word of caution: before upgrading, check on the word processor's capabilities for translating, converting, or importing existing word-processing files for use with the new software. In most cases this will not be a problem, but users of older, more obscure software should research this carefully.

Spreadsheets. This type of application software is used to perform mathematical, statistical, and financial calculations. The spreadsheet screen resembles an accountant's ledger sheet and is divided into vertical columns and horizontal rows. The columns and sheets are labeled alphabetically and the rows numerically. The intersection of rows and columns are known as cells. Depending on the program, a spreadsheet can contain anywhcre from thousands to millions of cells. Each cell can hold text, numbers, or formulas. Some of the more sophisticated spreadsheet programs add a third dimension—multiple or layered sheets.

The ability to apply formulas is one of the most valuable features of a spreadsheet. Formulas use data in other cells to calculate a result. With formulas in place, changes made to one cell can be reflected in calculations throughout an entire spreadsheet. In other words, a "ripple effect" takes place.

Spreadsheets are often used to prepare "what-if" analyses. These analyses instantaneously show the various outcomes when a change is made to a linked variable. A common what-if analysis is a loan amortization schedule. When the term or interest rate variable is changed, the spreadsheet automatically recalculates the payment amounts.

In MTS appraisal, spreadsheets are commonly used to prepare the asset listings that appear in the addenda of appraisal reports. Columns or categories of information that are often found in such asset listings include location, asset number, quantity, description, manufacturer, model, serial number, normal useful life, effective age, class, condition, date acquired, original cost, appraised value, photo number, comments, and pricing notes.

In an asset listing prepared using a spreadsheet, each row would contain a separate asset. By using a spreadsheet, as opposed to a word processor, the appraiser has the ability to manipulate the information in ways that expedite the appraisal process and create results that are consistent and accurate. Information can be manipulated in several ways including sorting, searching, rounding, and formatting.

Sorting refers to the ability to arrange data within a given range in an order that the user specifies. Data can be sorted in ascending or descending order. If an ascending sort order is selected, data are arranged alphabetically from *A* to *Z* and numerically from the smallest to the largest value. Conversely, if a descending sort order is selected, data is arranged in the opposite order (*Z* to *A*).

The process of sorting involves three fundamental steps. First, the entire range of information to be sorted is selected. Second, the column by which the range is to be sorted is chosen. Finally, the sort order (ascending or descending) is selected. If a secondary or multiple sort is desired, the second and third steps are repeated.

Sorting data can assist the MTS appraiser in several ways. For example, using a sample spreadsheet containing the columns for an asset listing mentioned above, how would an appraiser examine or value ali assets with a historical cost greater than $1,000? Without a spreadsheet program, the appraiser would have to search through the listing, asset by asset, page by page. By selecting the entire spreadsheet as the range, the historical cost as the sort column, and descending as the order, the data are arranged so that the assets with the highest cost would appear at the top and those with the lowest cost at the bottom of the spreadsheet. The appraiser can easily examine the assets meeting the $1,000 threshold.

Other columns or categories used for sorting include description, manufacturer, model, acquisition date, original cost, asset number, and classification. Sorting by description, manufacturer, or model results in like items falling together. This could make pricing faster and more consistent. Sorting by acquisition date and historical cost could assist in the application of trends or indexes. Sorting by asset number could streamline the process of verifying or reconciling assets, especially in instances where corresponding asset tags are in place. When using the sort feature, it is recommended that a "sort column" be added to the spreadsheet. This column should be created in advance of any sorting and contain sequential numbers that allow the spreadsheet to be restored to its original order.

Searching is another spreadsheet function that is extremely useful to the appraiser. This feature enables the appraiser to quickly locate specific information about an asset contained somewhere in a spreadsheet. Searching for an asset is a means of looking for a particular "string" or sequence of characters contained in one or more fields. As with sorting, the first step is to identify the range of information containing the item to be located. Next, the string of characters that pertains to the item to be found is typed. Using the previous sample spreadsheet, assume that a dealer responds to a request for information on a particular asset. When the request was made, the dealer's name and company were noted in the pricing notes column of the spreadsheet. Rather than painstakingly looking through a printout, the search function can be used to locate the dealer's name and the asset, and the appraiser can quickly record the pricing information.

Replacing is an additional option to the search function that can also help the appraiser work more efficiently. The replace option can change or update information on a range or spreadsheet-wide basis. For example, a client calls and says that he has provided erroneous information about his company's PC configurations. Instead of 1 MB of memory, they all have 4 MB of memory. By using the replace function, the entire spreadsheet can be searched for the string "1 MB" and replaced with the string "4 MB."

Rounding of values is routinely performed during the course of an appraisal. If the appraisal listing is quite large, a spreadsheet's rounding formula can save a significant amount of time. With this function comes the assurance that the appraised values are consistently rounded. Be aware, however, that rounding can affect calculations.

Formatting refers to the style or manner in which the contents of a spreadsheet cell are displayed and printed. Certain formats vary between cells containing alphanumeric information and cells containing numeric information. Common formats for both cell types include fonts, point size, bolding, italicizing, shading, underlining, outlining, and aligning. These formatting features enhance or emphasize information found in a spreadsheet. For cells containing numeric information, formatting can be used to display the figures as fixed numbers, with or without commas, or as currency. Numeric formatting also allows the display of a defined number of decimal places, bracketed, colored or minus sign (–) for negative numbers, and parentheses.

Originally, Lotus 1-2-3 (figure 9.3) was the most widely used spreadsheet program; however, Microsoft Excel is currently the top-ranking spreadsheet. Quattro Pro by Novell is a distant third. These three programs are nearly identical in terms of capability, functionality, and design. All provide DOS- and Windows-based versions, and for the most part, each program allows competing programs to open and exchange data files. Spreadsheet programs range in price from less than $100 for DOS-based versions to almost $400 for current, Windows-based versions. As with word processors and system software, special upgrade and competitive upgrade pricing are usually available.

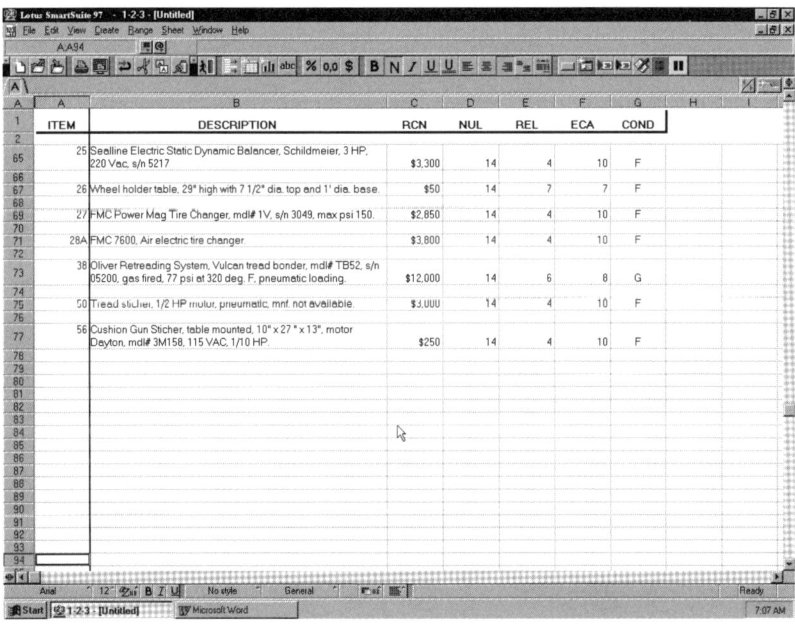

ITEM	DESCRIPTION	RCN	NUL	REL	ECA	COND
25	Sealline Electric Static Dynamic Balancer, Schildmeier, 3 HP, 220 Vac, s/n 5217	$3,300	14	4	10	F
26	Wheel holder table, 29" high with 7 1/2" dia. top and 1' dia. base.	$50	14	7	7	F
27	FMC Power Mag Tire Changer, mdl# 1V, s/n 3049, max psi 150.	$2,850	14	4	10	F
28A	FMC 7600, Air electric tire changer.	$3,800	14	4	10	F
38	Oliver Retreading System, Vulcan tread bonder, mdl# TB52, s/n 05200, gas fired, 77 psi at 320 deg. F, pneumatic loading.	$12,000	14	6	8	G
50	Tread sticher, 1/2 HP motor, pneumatic, mnf. not available.	$3,000	14	4	10	F
56	Cushion Gun Sticher, table mounted, 10" x 27 " x 13", motor Dayton, mdl# 3M158, 115 VAC, 1/10 HP.	$250	14	4	10	F

Figure 9.3. Sample Lotus spreadsheet.

Database Software

Database software refers to the programs used to store and retrieve data and is used to provide an organized collection of data. In most cases, databases are structured in terms of records and fields. A record is a particular listing or entry within a database. It usually contains several types of information that are known as fields. A record in a database might contain information about a specific person or company. A field is a particular category of information, such as a last name or a street address. There are several common types of fields including

- character fields (alphabetic information);
- numeric fields (numeric and monetary information);
- date fields; and
- time fields.

The American Society of Appraisers' directory is a good example of a database. Joe Smith, ASA, might be one record and Jane Doe, FASA, another. Fields might include discipline, name, company name, telephone number, fax number, and address.

There are two types of database programs: flat-file and relational. A *flat-file program* keeps all its information in a single file. Such programs are generally less sophisticated than relational databases, but they may be adequate for an appraiser's needs. Flat-file database programs frequently used by appraisers include PC-File, Professional File, Q&A, RapidFile, Reflex, and Superbase.

A *relational database* program stores information in various files. These files are linked together by key fields that contain matching information or values. Relational databases eliminate redundancy because information between files is shared rather than recreated. Relational database programs can be very large and powerful and are more efficient for complex uses. Relational database programs used frequently by appraisers include Access (figure 9.4), Approach, dBase, FoxPro, Rbase, and Paradox.

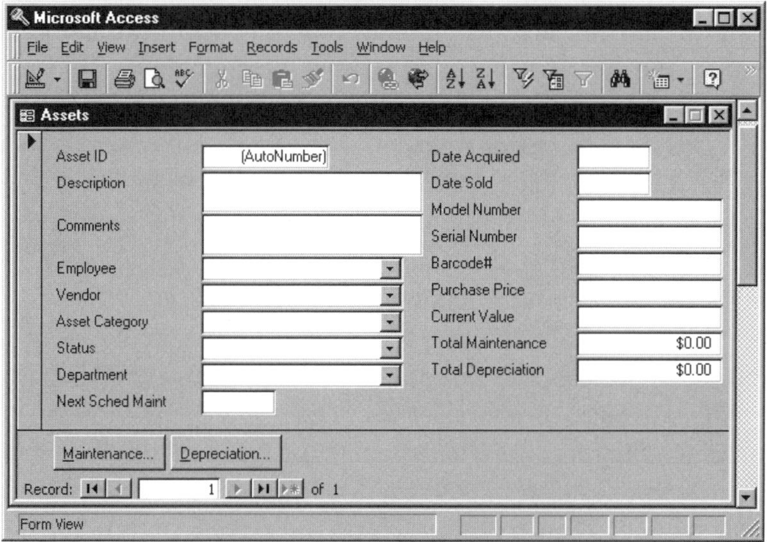

Fig. 9.4 Sample Access database.

Database programs can be used to perform many of the same tasks as spreadsheets. Sorting, searching, importing, and exporting of data are all routinely accomplished. Mathematical calculations, such as subtotaling, totaling, extending of values, and rounding of values, can be performed easily. The MTS appraiser can apply database software for several tasks, including large asset listings, client profiles, dealer and broker databases, and master pricing databases. Several such applications are described in more detail here.

A database is useful for generating and printing large asset listings. The number of assets that can be listed in a spreadsheet is limited to the number of rows the program contains. In Lotus 1-2-3, this number is currently 8,192. In most database programs, the number of assets or records is limited only by the storage capacity of the system's hard disk. In addition, the software can be programmed to automatically generate totals and subtotals when the data are printed. For instance, a simple program might be written to instruct the software to subtotal the fair market value column whenever the location field changes.

A database can keep track of *client profiles.* An appraiser might want to maintain information about clients, including general information such as client name, company, address, telephone number, and fax number. In addition, specific information including industry, date of last appraisal, conclusion(s) of value, purpose of the appraisal, contacts, and fee might be included.

Dealer and broker databases can track equipment dealers and brokers who have provided useful information for past appraisals. Data fields might include their specific areas of expertise, consulting rates, the engagements for which assistance was provided, and comment summaries.

Master pricing databases can consolidate pricing information for a specific category or class of assets. To create a master pricing database, an appraiser could combine all previously prepared asset listings for clients within the same industry. A medical-equipment master database, for instance, might include all the asset listings an appraiser has prepared for hospitals, medical clinics, treatment centers, and doctors' offices. In addition, it can include listings and prices from a host of other sources such as manufacturers, used equipment dealers, trade publications, classified advertisements, and pricing guides. By compiling this information in one master database, the appraiser has quick access to it and reduces valuation and research time by avoiding the proverbial reinventing of the wheel.

When selecting a database program, ensure that the software is compatible with any existing spreadsheet software. Most Windows-based databases are compatible with other Windows-based spreadsheets, but DOS-based programs have limited compatibility. Prices range from about $100 to more than $500 for the more sophisticated relational databases.

Selecting Application Software

Like hardware, software is continually changing. Each new version of system software or application software brings with it new features and enhancements. Hundreds of software companies and thousands of software programs are available, but Microsoft Corporation is the largest and most widely recognized. Its products come already installed on many new PCs.

It is important for the appraiser to use software that is both widely accepted and up-to-date. Although it is possible to purchase obscure software or older products for substantially less cost, the result may be incompatibility with the data of other appraisers and clients. When purchasing a new PC, check to see which software programs come "bundled" with the system. Frequently, PC manufacturers will include a host of software programs such as a word processor, a spreadsheet, and a database. These bundled programs are usually sufficient for most, if not all, of an appraiser's needs.

When purchasing or upgrading software for an existing PC, it is worth looking into manufacturer's packaged sets or suites, as they are sometimes called. Packaged sets include the three main applications and other applications such as presentation software and personal information managers. By purchasing a suite instead of individual programs, substantial savings can be realized. The most popular suites include Microsoft Office, which contains Word, Excel, Access, and PowerPoint; Lotus SmartSuite, which contains Word Pro, Lotus 1-2-3, Approach, and Freelance Graphics; and Corel Office Suite, with WordPerfect, Quattro Pro, Paradox, and Presentations. Prices in 1999 range from about $250 to $600.

Online Communications

Online communication refers to the ability to connect a PC to other computers via a modem on telephone lines using some type of communication software. Through online communications, a PC can access or log on to online services, the Internet, or other services.

Online Services

The term *online services* refers to the large commercial systems that users pay a subscription fee to access. These services provide information on a number of topics including news, business, sports, and weather. They also include features such as Internet access, e-mail capabilities, discussion forums, library centers, shopping areas, encyclopedias, databases, and other useful information. The major services include America Online, Prodigy, CompuServe, and Microsoft Network. These services provide their own communication software and direct access to the Internet.

It is advisable to subscribe to each service on a trial basis to explore the full range of services available. With this no-obligation, no-cost access, interested parties can test the service before they make a commitment. Subscription information and even free start-up software can be found in major computer publications.

Internet

The *Internet* is the term given to the International Network: a group of global information resources all linked to one another. Developed in the 1970s by the U.S. Department of Defense and originally called ARPAnet, the Internet has evolved to become an elaborate network of networks. (A network refers to two or more computers that are connected to one another.)

Accessing the Internet has become the most popular new phenomenon in the use of PCs. On the "Net," a smorgasbord of information is readily available. Networks can be found that include information on everything from government programs to unusual hobbies. Most of these sites are free, but this is changing as the Internet develops.

E-mail. One feature of the Internet frequently used is universal electronic mail, also known as e-mail. Electronic mail is a means of transferring data from one computer user to another. This data can be a letter or memorandum, digital image, spreadsheet file, a software program, or even a video clip. To use e-mail, a user establishes an e-mail address with an online service or a computer network. Once officially registered with an e-mail address, the user can begin sending and receiving information to anyone else with an e-mail address. An appraiser can use e-mail to send information to a friend on a commercial online service, a business colleague on a company network, or a client on the Internet. Today some appraisal reports are sent via e-mail instead of by conventional means such as the U.S. Postal Service or Federal Express. One caveat, however: the appraiser must be extremely cautious when transmitting appraisals by electronic means in order to protect appraiser-client confidentiality.

It is impossible to discuss the many different facets of the Internet in the context of this book. Currently the World Wide Web (WWW) is the area receiving the most attention. The WWW is a tool that uses hypertext to retrieve and display data and information on the Internet. Hypertext is a system whereby keywords are underlined or bolded and by selecting them users are linked to other related areas. The keyword serves as a launching point from which users travel to new but related topics or destinations. One of the fascinating aspects of the Web is the use of graphic images, photographs, and sound, and even video clips—not just simple text—to convey ideas. Individuals, businesses, governmental agencies, and institutions are all embracing the Web, creating Web pages or sites where others can access information about them. Each Web page has its own unique address or universal resource locator, commonly known as an URL.

Anther popular Internet feature is *Telnet,* which allows users to establish a connection with a remote computer. This way, two computer programs can work cooperatively by exchanging data over the Internet.

File transfer protocol (FTP) is a protocol that allows a user to transfer files among computers connected on the Internet. File transfers from a host computer are known as downloads while transfers to a host computer are known as uploads.

Client/server is a facility that allows a computer network to share its resources. Two distinct programs called the server and the client implement this sharing. The server provides a particular resource, and the client makes use of it. The most popular client/server facility is a service called "gopher." A gopher is a menu-driven server that executes requests for specific information. For example, a gopher server for a company might contain information useful to customers and employees such as product pricing or overtime schedules.

There are several ways to access the Internet. The online services previously described provide the easiest method. Other methods include Internet service providers (ISPs) and, more recently, telephone companies. ISPs are the least expensive (sometimes even free, with access through universities or from providers that derive their revenue from advertising), typically costing around $20 per month for unlimited access time in 1999. Typically, ISPs provide a local access telephone number, software, instructions, and support. The software used to connect to an ISP is commonly known as a Web browser. Two standard Web browsers are commonly used today: Netscape's Navigator and Microsoft's Internet Explorer. Telephone companies, including AT&T, MCI, and Pacific Bell, have recently entered the Internet arena, offering Internet access to their customers. Their pricing is comparable to ISPs at approximately $20 per month in 1999. They also provide software and local access telephone numbers.

Other Services

The uses of online communications for the MTS appraiser are endless. The appraiser can go online to obtain information on a particular company that is being appraised; collect data on an industry that is of interest solicit dealer price requests; access company price lists; search classified sections of major newspapers request information from other appraisers; and even seek employment. Appraisers now use online communications to post their qualifications and client lists on Web sites for access by potential clients. The appraiser, however, must exercise caution concerning client lists in order to protect appraiser-client confidentiality.

The American Society of Appraisers had an online service, the Appraisal Profession Online, which was replaced by a web site in 1998. The ASA web site URL is www.appraisers.org (see figure 9.5).

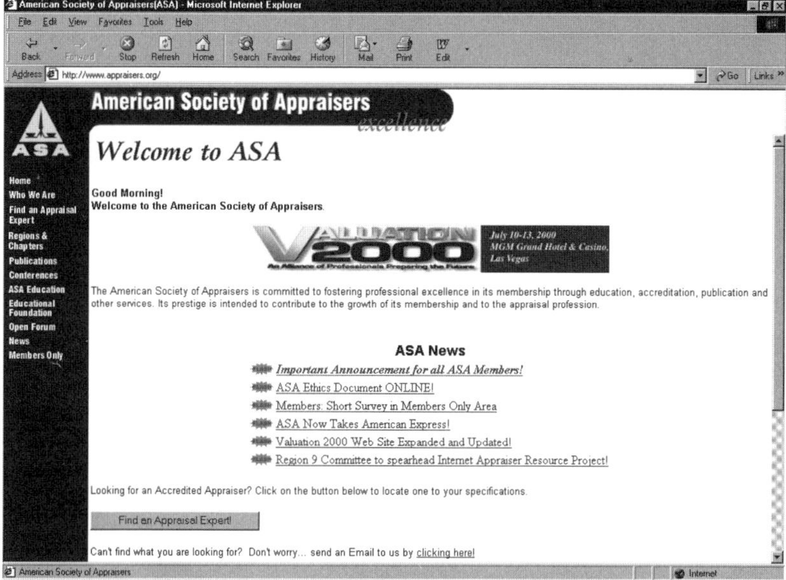

Figure 9.5.

Some other Internet sites that might be of particular interest to the MTS appraiser include the following:

Aircraft	
Aircraft Shopper Online	http://aso.solid.com/main4.htm

Antiques	
Antiques and the Arts Online	http://www.antiquesandthearts.com

Auto	
Car Guide	http://www.carprices.com/
Cars of Particular Interest	http://www.trader.com/vehicles
Glass's Car/Boat Guide	http://www.glassguide.com.au/
Kelley Blue Book	http://www.kbb.com/

Communications	
Broadband	http://www.broadband-guide.com

Computers	
Affiliated ReMarketing	http://www.remarketing.com/
Aice (computer memory)	http://www.aice.com/
American Computer Exchange	http://www.amcoex.com
American Computer Exchange	http://www.amcoex.com
American Teleprocessing	http://www.atchou.com/
Amitron	http://www.remarketing.com/broker/amitron
Arrowhead Resellers Corp.	http://www.remarketing.com/broker/arrow/
Ascent International	http://www.ascent.demon.co.uk/homepage.htm
Asset Remarketing Corp. (DG)	http://www.remarketing.com/broker/asset/
Beach Financial Corp (banking)	http://www.beachfinancial.com
Dynix Computer Services	http://www.interlog.com/~afford/
Spectra Company (mainframes)	http://www.Spectra911.com
Used Computer Mall	http://www.usedcomputer.com/

Construction	
Contractors Hotline	http://www.contractorshotline.com
GSN Construction Equipment	http://www.gsnet.com/

Financial	
Business Week	http://www.businessweek.com
Federal Government Information	http://www.fedworld.gov

Financial Data Finder	http://www.cob.ohio-state.edu/dept/fin/
Findex Financial Index	http://www.findex.com
Library of Congress	http://lcweb.loc.gov/homepage/lchp.html
Money Page	http://www.moneypage.com
The Thomas Register	http://www.thomasreg:ster.com
U.S. Treasury/IRS	http://www.irs.ustreas.gov/

Food Processing

Institute of Food Technology	http://www.ift.org/other.html

General

Asset Recovery Services	http://www.ars-equipment.com/
Chicago Tribune	http://www.chicago.tribune.com
Classified Flea Market	http://www.cfm.com/
New York Times	http://www.nytimesfax.com
SF Chronicle Classifieds	http://www.sfna.com/cgi-bin/main.pl
Surplus Governmental Sales	http://www.drms.dla.mil/
Surplus Record	http://www.surplusrecord.com/
The Industrial Group	http://www.industrialgroup.com
Used Equipment Network	http://buyused.hsix.com/

High Technology

Dovebud	http://www.dovebud.com
Semiconductor Handilinks	http://www.ahandyguide.com
Technology Network	http://www.tecnet.com/tecnet/Contract.htm
Used Semiconductor Dealers	http://www.big-list.com/
Aaron Process Equipment	http://www.aaronequip.com/
Federal Equipment Co. (Process)	http://www.fedequip.com/

Laboratory Equipment

Lab Equipment Exchange	http://www.labx.com/
LabTrader	http://www.labtrader.com/

Machine Tools

The Machine Tool Exchange	http://www.machtools.com

Maritime	
Intership Ltd. Barges-Vessels	http://www.barges.com
Medical Equipment	
Mednet	http://www.mednetlocator.com
Pemed Medical Equipment	http://www.pemed.com/
The Medical Marketplace	http://www.mkpl.com/cmp/mmp/mmphome.html
Used Equipment World (Medical)	http://www.uew.com/indexct.htm
Organizations	
American Society of Appraisers	http://www.apo.com
American Society of Appraisers Northern California	http://www.appraisers-norcal.com
Appraisal Institute National Commercial	
Real Estate Database	http://www.realworks.com
Personal Property	
Virtual Library Museums	http://www.comlab.ox.ac.uk/archive/other/museums.html
Photography	
Camera Traders Guide	http://www.cameratradersltd.com
Online Camera Exchange	http://www.tfb.com/oce/oce.htm
Photography	http://www.photoshopper.com
Plastics	
Plastics Hotline	http://www.plasticsnet.com/
Printing	
Worldwide Printing Equipment	http://printequip.com/
Radio and Television	
Broadcast Exchange	http://www.broadcastexchange.com
Real Estate	
Mortgage Market Information	http://www.interest.com
Realty Blue Book	http://www.deheer.com/bluebook/home.shtml
Reference	

Telephone Numbers	http://www.switchboard.com
The Ultimate Yellow Pages	http://www.theultimates.com
World Wide Yellow Pages	http://www.yellow.com/
Testing Equipment	
Metric Equipment Sales	http://www.metricsales.com
The Test Equipment Depot	http://www.fotronic.com/
Textiles	
Textile Red Book	http://www.atimag.com/trb.html
Travel	
ESRI (maps)	http://www.esri.com/company/free.html
Hotels and Travel	http://www.hotelstravel.com
Yahoo Maps	http://www.mapblast.com/myblast/index.mb
Trucks	
Big Truck Trader	http://www.trucktraderonline.com
Orix Credit Alliance	http://www.orix-cac.com
Woodworking	
Ex Factory	http://www.exfactory.com/
Woodworking Catalog	http://www.woodworking.com/index.html

Key Points

- Personal computers are an indispensable tool in MTS appraisal. It would be difficult to imagine preparing an appraisal without them. They perform many of the necessary tasks needed to prepare appraisals in a timely manner, a requisite of today's faster world.

- In 1999, a well-configured personal computer would include a Pentium CPU, 64 MB of RAM, a ten-gigabyte hard drive, 3.5" floppy drive, CD-ROM drive, sound card, keyboard, mouse, 56 BPS modem, SVGA card, 56k modem, and 0.28 SVGA monitor.

- Software would include system software such as Windows, a word processor such as Microsoft Word or WordPerfect, a spreadsheet such as Lotus 1-2-3 or Excel, a database such as Access or dBase, and communication software for online communication with online services and Internet access.

- Though PCs can simplify the appraisal process, they should be considered tools that facilitate the application of sound appraisal research and technique. They should be used in a manner that enhances the quality of the work product, not to shortcut or bypass the steps necessary to produce documented appraisal findings.

Additional Reading

Bouton, Barbara, ed. *World Wide Web Yellow Pages.* 1st ed. Indianapolis: New Riders Publishing, 1995.

Compton's Interactive Encyclopedia CD. City, state: Computer Technology, 1995.

Doyle, Casey D., ed. *Computer Dictionary.* 2nd ed. Redmond, WA: Microsoft Press, 1994.

Hahn, Harley, and Rick Stout. *The Internet Complete Reference.* Berkeley, CA: Osborne McGraw-Hill, 1994.

Microsoft Encarta Encyclopedia CD. Redmond, WA: Microsoft Corporation, 1995.

O'Mara, Michael, and Gerry Routledge. *Using Windows 95.* 1st ed. Carmel, IN: Que Corporation, 1996.

O'Mara, Michael, and Gerry Routledge. *Using Your Personal Computer.* 2nd ed. Carmel: IN: Que Corporation, 1996.

Case Study

Today's MTS appraiser uses the PC in many facets of the job. Following is an example of how an appraiser can use some of the components and software mentioned throughout this chapter for a typical engagement that begins with a call from a potential client, XYZ Corporation.

John, XYZ's controller, has called Joe Smith of J.S. Appraisal Company with an interest in having his company's furniture, fixtures, and equipment appraised. John requests Joe's qualifications and information about J.S. Appraisal Company. Joe uses his *word processing software* to prepare a cover letter and print his qualifications. Sensing some urgency in John's request, Joe decides to send the information via fax using his PC's *fax-modem*.

John is impressed by Joe's credentials and his timely response and awards J.S. Appraisal Company the assignment. As it turns out, the XYZ Corporation has extremely accurate fixed-asset records that are maintained on the company's mainframe computer. Joe realizes that these records could be quite useful in the appraisal process and requests that John have the information downloaded from the mainframe to a PC-based *spreadsheet* file, which is saved on a disk.

Once Joe receives the disk, he uses his spreadsheet program on his notebook PC to open the file and sort the information by department and by cost. He then formats and prints the information in a manner that allows him to efficiently perform his field inspection. He inserts a blank row between assets, creating space to add descriptive information, and adds several blank columns for tag numbers and condition ratings.

After concluding his inspection, Joe returns to his office and incorporates the new information into the spreadsheet file. He deletes the assets that were not found and adds assets that were not on the list.

Joe then begins his valuation process. By sorting the spreadsheet in order of descending cost, Joe determines that only 20 assets make up 90 percent of the company's asset expenditures. He concludes that his valuation efforts would be most efficient if he spent most of his time on these assets. (This does not mean that Joe will ignore the remaining assets.)

Before he begins contacting dealers and manufacturers, Joe starts his *database program* and opens his master pricing database file that contains all the major assets he has valued over the past five years. He vaguely recalls having priced similar assets some time ago but cannot remember the specific job. A search of his database file brings up several similar assets and directs him to the specific job name. Upon retrieving the file, Joe finds a list of sources he had previously contacted and uses this information as a starting point.

One of the sources now furnishes a price list on the World Wide Web. Using an *online service,* Joe accesses the company's *Internet* URL and downloads the current price list.

Joe completes his pricing research and incorporates his findings in a new column he has created in the spreadsheet, which in this case is entitled "FMV" for fair market value. Joe does another sort by department and classification, as requested by John, and then inserts the proper formulas to subtotal and total the values. He prints the spreadsheet entitled "asset detail" for incorporation into his narrative report.

Joe then opens his word processing software and brings up a report *template* that prompts him to enter the unique information pertaining to his appraisal for XYZ Corporation. After completing the report, he opens an invoice template and prepares his bill. Along with his report, he sends John a disk containing the asset detail spreadsheet. John is so impressed that he recommends Joe to his company's major clients.

Before closing out the job, Joe imports the pertinent information from the asset detail spreadsheet into his master pricing database file. He also adds John to his client profile database and XYZ Corporation to a word-processing document entitled "representative clients."

10

Technical Specialties

Objectives:

1. Define technical specialties.
2. State types of technical specialties.
3. Define the economic concepts affecting the appraisal.
4. State the best appraisal approach for each specialty.
5. Recognize the typical client for each specialty.

The ASA consolidated the machinery and equipment discipline and technical valuation discipline into the machinery and technical specialties discipline in 1994. While machinery and equipment appraisers are professionally qualified to value most types of machinery and equipment used in industrial and commercial uses, they may not have the expertise required to value certain types of assets without further training and experience. That training includes additional continuing education, working with experts in a specialty, and personally conducting research. The competency provision of USPAP allows an appraiser to accept an appraisal in a specialty without prior experience only after explaining the lack of experience to the client and then doing the research necessary to become competent to perform the appraisal or by enlisting the services of a qualified specialist.

Technical specialties appraisers are specialists who have studied and concentrated in areas that require unique training and specific experience in a particular industry, category of property, or discipline. Each specialty requires an understanding of a particular body of knowledge before a valuation study, analysis, or appraisal can be properly performed. Each specialty addresses certain types of assets in a manner consistent with its industry or discipline, and each appraisal assignment addresses different economic concepts and methods that affect the value of those assets and their users. This chapter describes the unique specialties within the MTS discipline and discusses the types of assets valued within the specialties, the economic concepts that affect value, and who uses these specialty appraisal services. All three valuation methods (cost, sales comparison, and income approaches) are investigated here, but only the most commonly used indicators are discussed for each specialty. It is important to remember that USPAP requires that all three approaches be considered, although not all may ultimately be used.

Aircraft Specialty

The aircraft specialist typically appraises most types of aircraft, support equipment, associated spare parts, and inventory. Typical assets include the smallest Cessna 150 single-engine plane to the largest Boeing 747 commercial aircraft, as well as helicopters and training craft, including corporate, commuter, and private reciprocating engine and jet aircraft. Aviation support equipment, such as spare parts, special tools, and ground-support equipment, is also included.

Economic concepts that influence aircraft values include regulations concerning noise and air pollution, the U.S. and world economies, passenger trends, and airfare and fuel costs. Because fatigue and corrosion can be major concerns, especially on aging aircraft, an important part of an aircraft appraisal is the property inspection. The appraiser must be familiar with federal aviation regulations (FARs) concerning maintenance schedules, documentation, and approved replacement parts that are used on the aircraft being appraised. Federal Aviation Administration (FAA) requirements may vary with the types and uses of aircraft. FARs and fleet directives are innumerable and must be investigated in every appraisal. It is important to match the current published regulations and directives with the aircraft being appraised and to carefully inspect the aircraft to ensure that it is in full compliance.

Although the income and cost approaches to value are considered, the strongest evidence of value usually comes from the sales-comparison approach. It may be difficult to verify some details of a sale, but the sales-comparison approach is usually relied upon as the most meaningful valuation indicator. The most common values required are fair market value or some form of liquidation value, either as of a current date or a date in the future. Clients requesting aircraft appraisals include aircraft owners (airlines, corporations, financial institutions, and individuals), lenders, and lessors.

Cost Survey Specialty

Appraisers who hold this specialty designation are involved mainly in the allocation of value or cost among the various components of real property and personal property in industrial operations, factories, office buildings, and other commercial or industrial settings. They are experts in the cost method of valuation as required for insurance coverage or insured loss analysis; accounting and tax allocation in mergers, acquisitions, and other forms of business sales; cost segregations or depreciation lifing studies for income tax purposes; and valuation or cost analysis in connection with various environmental and other tax credits.

The cost approach is the most meaningful indicator used by the cost survey specialist. The cost survey specialist is familiar with sources of current cost data; indexes; and quantity survey, unit-in-place, and other methods of estimating costs. Physical depreciation and functional obsolescence are measured by observation, age/life methods, and other standard procedures such as cost to cure. Cost survey assignments often require a working knowledge of the various facets of the construction industry, ranging from tract homes to high-rise commercial buildings to complex industrial structures.

Other services provided by these specialists, in addition to those already mentioned, include but are not limited to cost analysis for environmental or other tax credits; quantitative and qualitative analysis of historical buildings and structures; cost monitoring of construction projects where progress payments are made by lenders; analysis of building losses from man-made or natural disasters; verification and confirmation of public funding for governmental projects; and feasibility analysis of proposed projects. Typical clients for cost surveys are commercial and industrial property owners, government agencies, lending institutions, and insurance companies.

Industrials Specialty

The industrials specialist values the property, plant, and equipment of heavy industry and the process industries, including but not limited to, petrochemical plants, oil refineries (figure 10.1), steel mills, paper mills, food-processing plants, and similar facilities.

Economic concepts applicable to the industrials specialty may involve global, national, regional, or local economic activity, including plant location and environmental regulations. All three approaches to value are frequently used to develop an opinion of value. Various forms of obsolescence in the cost approach are frequently analyzed. Excess capital costs and excess operating cost penalties are often quantified; therefore, the ability to develop costs for state-of-the-art facilities for the industry in question is important. The income approach, particularly the discounted cash-flow approach, is commonly used by these specialists. Hence, familiarity with business enterprise analysis and financial valuation techniques is required. The sales comparison approach is used if meaningful data can be obtained concerning sales of comparable properties.

Appraisal services of this type are typically requested by large industrial property owners for purposes as varied as property tax appeals, allocation of purchase price for book and income tax, insurance, and financing. Skills needed include a detailed knowledge of the industry in question, experience in the three valuation approaches, and the ability to testify in court.

Figure 10.1. Oil refinery. ***Courtesy of Total Petroleum***

Marine Survey Specialty

Marine survey specialists have two areas of specialization: commercial marine and yachts. Some appraisers are qualified in both. Typical yachts appraised include production model and custom powerboats, sailboats, and houseboats. Commercial marine vessels appraised include excursion, dinner, or gambling vessels; dredges; dry docks and docking facilities; drill rigs; and all types of standard and specialized ocean commerce vessels.

The sales-comparison approach is the most common in yacht valuation. Comparable sales are examined and adjusted based mainly on age, condition, and location adjustments; sometimes the time of year is an important factor. The cost approach is given primary consideration with rare or custom boats. Commercial-vessel valuations usually start with the cost approach, with adjustments being made for all forms of accrued depreciation, including condition, remaining useful life, suitability for service, certification or lack thereof, maintenance history, location, and relocation costs. Traditionally the income approach has not been used in this specialty, as proprietary financial information is generally not readily available.

These specialists typically provide such services as

- damage surveys and appraisals, usually required by insurance companies, in which descriptions of the nature and extent of damage are detailed and repairs to restore property to its prior condition are recommended; and
- surveys or appraisals to determine condition and value upon sale of yachts, small craft, and ships.

Clients for marine appraisals are sellers or prospective purchasers, those with insurance interests, financial institutions, attorneys, and government agencies.

Natural Resources, Mines and Quarries, and Oil and Gas Specialties

Specialists in natural resources, mines and quarries, and oil and gas appraise assets such as national forests; mineral, oil, and gas reserves; and water rights. Natural resources include forestlands and water rights. Mineral reserves include coal, limestone, sand and gravel, gold, lead, and other hard minerals. Oil and gas reserves include hydrocarbon-based assets of a tangible and intangible nature in a producing or unproven field.

Economic concepts that affect value in these specialties include the total recoverable quantity, production rate, and remaining life of the reserves or resources being appraised. Current and projected future selling prices of reserves or resources significantly influence value. Value is also influenced by the global market, supply and demand, and market rationalizations. In addition, changing federal tax laws, investor-required returns-on-investments, and the cost of debt affect value. Typical approaches to value include the sales comparison and income approaches. The income approach is usually developed using discounted cash-flow analysis, income capitalization, or the capitalization of royalties. The sales-comparison approach is used primarily for surface assets. The cost approach, while considered, is typically not used because of the nature of the items being appraised.

Typical clients of this specialty are government agencies, mining companies, financial institutions, taxation agencies, attorneys, industrial companies, and individuals.

Public Utility and Railroad Specialties

Public utility and railroad specialists appraise assets owned by companies that are regulated by state or federal governments (although many of these companies are becoming deregulated). This specialty requires special knowledge to take into account the unique economic and value characteristics of public utility and railroad properties (figure 10.2) and to properly recognize regulatory factors that influence value conclusions. Typical users of these appraisal services include electric and telecommunications utilities; gas distribution and transmission companies; cable and satellite systems; airlines; water and sewer systems; and transportation systems. Railroad specialists appraise assets used in the railroad industry which, like public utilities, is regulated but becoming less so.

Figure 10.2. Photograph of railroad scene. *Courtesy of Union Tank Car Company*

Economic concepts affecting value within these industries include federal and state regulations, the general level of economic activity, the stock and bond markets, and national and local debt costs. Utilities have special accounting requirements and the appraiser must be familiar with the applicable uniform system of accounts.

Approaches to value include all three approaches: cost, sales comparison, and income. The cost approach is frequently used for unique types of assets for which there is no quantifiable income stream and no reliable market or sales data.

For railroad and utility property, a unit valuation type of approach is commonly used to develop a value for the entire business enterprise or going concern. The unit valuation approach is a technique that uses parts of all three indicators of value at the same time. Because of rate-base regulation and the emphasis on strict property accounting rules, the cost approach is typically based on the inventory represented by the property records and its historical costs, which are often adjusted to value using cost trends and depreciation techniques based on age. The income approach to value is often given considerable weight.

These specialists estimate values, both tangible and intangible, for properties in connection with rate-case studies, sale or acquisition, eminent domain (condemnation), property tax appeals, and insurance placement.

Other Specialties

Other specialists within the MTS discipline include agricultural chattels, arboriculture, and computers and high-tech personal property.

Agricultural Chattels

Specializing in crops, livestock, agricultural vehicles, farm and ranch machinery and equipment, and other non-real-estate assets, the agricultural chattels specialist is concerned with both fair market value and various definitions of liquidation value for collateralization, estates, property sales, and futures markets. These specialists must also be cognizant of real property appraisal when dealing with crops that are real estate until harvested but then become personalty.

Arboriculture

The arboriculture specialist values trees and other plantings. Some of the arboricultural appraiser's assignments involve condemnation cases, establishing loss for income tax purposes, settlement of insurance claims relating to loss or destruction of trees and other landscape plantings, and horticultural/forensic testimony in court cases.

Computers and High-Tech Personal Property

These specialists appraise all types of computers, ranging from small PCs to the largest mainframes, and all of the related components, including tape and disk drives and printers. They also appraise complex electronic equipment, including everything from large telephone switching systems to oscilloscopes, from medical and dental equipment to small handheld instruments to X-ray machines and CAT scanners.

Key Points

- Technical specialties appraisers are involved in the appraisal of aircraft, industrial properties, marine equipment, public utilities, railroads, agricultural chattels, arboriculture, and computers and high-tech equipment, as well as perform cost surveys.
- Like other professions, the machinery and equipment appraisal profession has grown to encompass fields of specialization. To be successful in these fields, the appraiser must develop a concentrated knowledge, obtain additional training, and be familiar with the applicable valuation methods.

Technical Specialties References

The Airliner Price Guide of Commercial-Regional and Commuter Aircraft. Airliner Price Guide, Inc. Oklahoma City, OK

Center for Management Development, Wichita State University. "Appraisal: Communication, Energy and Transportation Properties for Ad Valorem Taxation" (conference proceedings).

Aircraft Bluebook-Price Digest. City: Intertec Publishing Corporation Cost Survey Specialty. Oklahoma City, OK.: Aircraft Bluebook Corp.

BUC'S Used Boat Price Guides. BUC Research. 1st issue 1961 Fort Lauderdale, FL.

Chemical Marketing Reporter. Schnell Publishing Company. New York (began Oct 1979; ceased April 1994).

Fairplay: The International Shipping Weekly. Prime Publications. London, Financial Times.

Hydrocarbon Processing. Gulf Publishing Company. Houston, TX..

Means Construction Cost Data Books. Kingston, MA: R. S. Means Company, Inc., annual (current ed. is 57th).

Mines and Quarries Engineering and Mining Journal. Maclean Hunter Publications. Toronto, Ontario, Canada.

NADA Used Boat Guide. McLean, VA, National Automobile Dealers Association.

Oil and Gas Journal. PennWell Publishing Company. Nashua, NH (magazine).

Progressive Railroading. Trade Press Publishing Corporation. Milwaukee, WI (magazine).

Public Utilities Fortnightly. Public Utilities Reports. Vienna, VA.

Railway Age. Simmons-Boardman Publishing Corporation. Editor, Luther S. Miller New York

Richardson Construction Estimation. Richardson Engineering Services. Mesa, AZ.

U.S. Bureau of Mines (source of data), Department of the Interior. Washington, DC.

Workboat: In Business on the Coastal and Inland Waterways. Journal Publications.

Appendix A

Accounting Depreciation

What follows is a simple overview of accounting methods and terminology. It is included primarily to demonstrate that accounting depreciation differs significantly from appraisal depreciation.

In calculating depreciation, companies typically use one of the following four methods:

- straight line,
- units of activity,
- double declining balance, and
- sum-of-the-years digits.

Straight-line depreciation calculations are a constant amount for each period. It is the depreciable cost divided by the asset's useful life. The formula that gives the amount of depreciation to be taken each year is

$$D = C/n,$$

where D equals the annual depreciation amount, C equals the depreciable (generally not including salvage value) cost, and n equals the depreciable life in years. Assume for example that an asset has a five-year life with an initial cost of $11,000 and a salvage value of $1,000. The depreciable cost would be $10,000, and the annual depreciation rate would be 20 percent (100 percent over five years). The depreciation amounts are shown in table A.1.

Year	Depreciable Cost		Depreciation Rate		Annual Depreciation	Accumulated Depreciation	Net Book Value
1	* $ 10,000	x	20%	=	$2,000	$2,000	$9,000
2	$ 10,000	x	20%	=	$2,000	$4,000	$7,000
3	$ 10,000	x	20%	=	$2,000	$6,000	$5,000
4	$ 10,000	x	20%	=	$2,000	$8,000	$3,000
5	$ 10,000	x	20%	=	$2,000	$10,000	$1,000

* *Note: The annual depreciation is subtracted from the initial cost [$11,000 – $2,000 = $9,000]. Note that this is for calculating accounting depreciation only. The salvage value used in this method is not usually considered in appraisals.*

Table A.1. Straight-Line Depreciation

Units of activity is similar to straight line; however, the depreciation calculation is based on hours used or some other production unit and may vary from the current accounting period to the next period. As an example, suppose a machine has a life of 100,000 hours and its initial cost is $11,000 with a salvage value of $1,000 and a depreciable cost of $10,000. The depreciation cost per unit is calculated by dividing the depreciable cost by the total life. In this case the depreciation cost per unit is $10,000/ 100,000 hours which equals $0.10 per unit. The depreciation, assuming the given annual hours used, would be calculated for each year as shown in table A.2.

Year	Hours		Depreciation Cost/Unit		Annual Depreciation	Accumulated Depreciation	Net Book Value
1	20,000	x	$0.10	=	$2,000	$2,000	$8,000
2	40,000	x	$0.10	=	$4,000	$6,000	$6,000
3	15,000	x	$0.10	=	$1,500	$7,500	$2,500
4	15,000	x	$0.10	=	$1,500	$9,000	$1,500
5	10,000	x	$0.10	=	$1,000	$10,000	$1,000

Table A.2. Units of Activity

The *double declining balance* depreciation method is an *accelerated* depreciation method because it results in a higher depreciation in the early years. In this method, the depreciation rate remains constant while the book value to which the rate is applied is declining each year. Under this method, the salvage value is not considered until the cost is fully depreciated. This method can be applied to the previous straight-line example: an asset with a five-year life, initial cost of $11,000, and a salvage value of $1,000. The depreciable cost would still be $10,000. The annual depreciation rate would be twice the straight-line rate of 20 percent for a total of 40 percent. The depreciation amounts are calculated as shown in table A.3.

Year	Beginning Net Book Value		Depreciation Rate		Annual Depreciation	Accumulated Depreciation	Net Book Value
1	$ 11,000	x	40%	=	$4,400	$4,400	$6,600
2	$ 6,600	x	40%	=	$2,640	$7,040	$3,960
3	$ 3,960	x	40%	=	$1,584	$8,624	$2,376
4	$ 2,376	x	40%	=	$950	$9,574	$1,426
*5	$ 1,426	x	$1,000	=	$426	$10,000	$1,000

**Note:* In Year 5 the salvage value of $1,000 limits the amount to $426.*

Table A.3. Double Declining Balance.

The *sum-of-the-years digits* method is also an accelerated depreciation method. In this method, a declining fraction is applied to the depreciable cost. The fraction's numerator is equal to the remaining service at the beginning of each year, and its denominator is equal to the sum of the total years over the estimated useful life. For a five-year life the denominator is equal to $(1 + 2 + 3 + 4 + 5) = 15$. The calculations [assuming a salvage value of $1,000] are demonstrated in table A.4.

Year	Depreciable Cost		Depreciation Rate		Annual Depreciation	Accumulated Depreciation	Net Book Value
1	* $10,000	x	10/55 = .18	=	$1,800	$1,800	$9,200
2	$10,000	x	9/55 = .16	=	$1,600	$3,400	$7,600
3	$10,000	x	8/55 = .15	=	$1,500	$4,900	$6,100
4	$10,000	x	7/55 = .13	=	$1,300	$6,200	$4,800
5	$10,000	x	6/55 = .11	=	$1,100	$7,300	$3,700
6	$10,000	x	5/55 = .09	=	$900	$8,200	$2,800
7	$10,000	x	4/55 = .07	=	$700	$8,900	$2,100
8	$10,000	x	3/55 = .05	=	$500	$9,400	$1,600
9	$10,000	x	2/55 = .04	=	$400	$9,800	$1,200
10	$10,000	x	1/55 = .02	=	$200	$10,000	* $1,000

*Note: *Assumes salvage value is $1,000.*

Table A.4. Sum-of-the-Year Digits

Appendix B

Compound Interest

Appraisers should understand the theories of compound interest and present value and the latter's application in calculating various obsolescence penalties. Compound interest is calculated on an ever-increasing basic sum. For example, if 10 percent compound interest were applied to $1.00 for a period of five years, the total principal and interest at the end of the five-year period would be $1.6105 as shown in table B.1.

Year	Interest Calculations	Accumulation of Interest	Accumulation of Principal and Interest
0	Initial Amount		$1.00
1	10% x $1.00 = $0.10	$0.10	$1.10
2	10% x $1.10 = $0.11	$ 0.11	$1.21
3	10% x $1.21 = $0.121	$ 0.121	$1.331
4	10% x $1.331 = $0.1331	$0.1331	$1.4641
5	10% x $1.4641 = $0.1464	$0.1464	$1.6105
Totals		$0.6105	$1.6105

Table B.1 Compound Interest

The above calculations can be checked by referring to appendix C, table C.3; under the 10 percent column and across the five-year row is the factor of 1.6105, which, when multiplied by $1.00, equals $1.6105.

Interest Calculations

Modern calculators and computer spreadsheet programs can simplify compound interest and time value of money calculations. Alternatively, tables providing the numerical factors used in compound interest calculations are available in most financial or mathematical texts. Four such tables are included in appendix C:

Table C.1. Present Value of $1.00
Table C.2. Present Value of an Annuity of $1.00 per Period
Table C.3. Future Value of $1.00 at the End of *n* Periods
Table C.4. Future Value of $1.00 per Period for *n* Periods (Sum of an Annuity of $1.00 per Period)

Table C.4 is similar to table C.3, except that C.4 is for an annuity requiring payments every year while C.3 is for one initial payment. For example, an annuity based on 10 percent interest and $1.00 per year would be calculated as shown in table B.2.

Year	Interest and Payment Calculation	Annual Payments	Accumulation of Principal and Interest
1	Initial Amount	+ $1.00	$ 1.00
2	10% x $1.00 = $0.10	+ $1.00	$ 2.10
3	10% x $2.10 = $0.21	+ $1.00	$ 3.31
4	10% x $3.31 = $0.331	+ $1.00	$ 4.641
5	10% x $4.641 = $0.4641	+ $1.00	$ 6.1051
Totals		+ $5.00	$ 6.1051

Table B.2. Interest Calculations

Check these calculations against table C.4; using the 10 percent interest column and five-year row, the factor is 6.1051, which when multiplied by $1.00 yields $6.1051.

The concept of present value (it is sometimes called present worth) is basically the inverse of compound interest. Stated another way, a dollar today is worth more than a dollar in the future. If one invests 91 cents today at 10 percent interest, it will grow to one dollar in one year. Thus, the present value of $1.00

to be received one year in the future is 91 cents today at 10 percent interest. To receive $1.00 per year during each of the next two years, how much would have to be invested today? At 10 percent and considering each year separately, the investment should be 91 cents today for the first year and approximately 82.6 cents today to receive $1.00 for each of the next two years. Therefore, the present value of $1.00 to be received in each of the next two years is 91 cents plus 82.6 cents or $1.736 at 10 percent. This is the process of discounting in its simplest form. It is the conversion of expected future periodic payments to present value.[1]

Present value factors can be manually calculated as shown below. Again, assume a 10 percent interest rate, $1.00 as principal, and a five-year term. One dollar discounted in the first year would be represented by $0.9091, derived by dividing $1.00 by $1.10. The results of the five-year discounting are reflected as follows:

Year 1	=	$ 1.0000 ÷ 1.10	=	$0.9091
Year 2	=	$ 0.9091 ÷ 1.10	=	$0.8264
Year 3	=	$ 0.8264 ÷ 1.10	=	$0.7513
Year 4	=	$ 0.7513 ÷ 1.10	=	$0.6830
Year 5	=	$ 0.6830 ÷ 1.10	=	$0.6209

The calculation is graphically displayed in table B.3.

		$1	$1	$1	$1	$1
Years (Periods)	0	1	2	3	4	5
1	$ 0.9091					
2	$ 0.8624					
3	$ 0.7513					
4	$ 0.6830					
5	$ 0.6209					
Total	$ 3.7907					

Table B.3. Present Value Graph

The above calculation can be verified by examining table C.1 at the 10 percent column and five-year row amount.

The present value concept can be used to assist with estimating various forms of functional and economic obsolescence. For example, as discussed in chapter 3, present value calculations are used to quantify functional obsolescence caused by excess operating costs because the excess costs usually occur over a term of years.

Notes

[1] Robert S. Svoboda. 1989. "Fair Market Value Concepts." In *Appraising Machinery and Equipment*, p. 99.

Appendix C

Financial Tables

The use and application of the following financial tables was covered in appendix B. The following tables are included here:

Table C.1. Present Value of $1.00

Table C.2. Present Value of an Annuity of $1.00 per Period

Table C.3. Future Value of $1.00 at the End of *n* Periods

Table C.4. Future Value of $1.00 per Period for *n* Periods (Sum of an Annuity of $1.00 per Period)

Present Value of $1

$$PV = \frac{1}{(1+k)^n}$$ where "k" = rate and "n" = number of periods.

Period	1%	2%	3%	4%	5%	6%	7%	8%	9%
1	0.9901	0.9804	0.9709	0.9615	0.9524	0.9434	0.9346	0.9259	0.9174
2	0.9803	0.9612	0.9426	0.9246	0.9070	0.8900	0.8734	0.8573	0.8417
3	0.9706	0.9423	0.9151	0.8890	0.8638	0.8396	0.8163	0.7938	0.7722
4	0.9610	0.9238	0.8885	0.8548	0.8227	0.7921	0.7629	0.7350	0.7084
5	0.9515	0.9057	0.8626	0.8219	0.7835	0.7473	0.7130	0.6806	0.6499
6	0.9420	0.8880	0.8375	0.7903	0.7462	0.7050	0.6663	0.6302	0.5963
7	0.9327	0.8706	0.8131	0.7599	0.7107	0.6651	0.6227	0.5835	0.5470
8	0.9235	0.8535	0.7894	0.7307	0.6768	0.6274	0.5820	0.5403	0.5019
9	0.9143	0.8368	0.7664	0.7026	0.6446	0.5919	0.5439	0.5002	0.4604
10	0.9053	0.8203	0.7441	0.6756	0.6139	0.5584	0.5083	0.4632	0.4224
11	0.8963	0.8043	0.7224	0.6496	0.5847	0.5268	0.4751	0.4289	0.3875
12	0.8874	0.7885	0.7014	0.6246	0.5568	0.4970	0.4440	0.3971	0.3555
13	0.8787	0.7730	0.6810	0.6006	0.5303	0.4688	0.4150	0.3677	0.3262
14	0.8700	0.7579	0.6611	0.5775	0.5051	0.4423	0.3878	0.3405	0.2992
15	0.8613	0.7430	0.6419	0.5553	0.4810	0.4173	0.3624	0.3152	0.2745
16	0.8528	0.7284	0.6232	0.5339	0.4581	0.3936	0.3387	0.2919	0.2519
17	0.8444	0.7142	0.6050	0.5134	0.4363	0.3714	0.3166	0.2703	0.2311
18	0.8360	0.7002	0.5874	0.4936	0.4155	0.3503	0.2959	0.2502	0.2120
19	0.8277	0.6864	0.5703	0.4746	0.3957	0.3305	0.2765	0.2317	0.1945
20	0.8195	0.6730	0.5537	0.4564	0.3769	0.3118	0.2584	0.2145	0.1784
21	0.8114	0.6598	0.5375	0.4388	0.3589	0.2942	0.2415	0.1987	0.1637
22	0.8034	0.6468	0.5219	0.4220	0.3418	0.2775	0.2257	0.1839	0.1502
23	0.7954	0.6342	0.5067	0.4057	0.3256	0.2618	0.2109	0.1703	0.1378
24	0.7876	0.6217	0.4919	0.3901	0.3101	0.2470	0.1971	0.1577	0.1264
25	0.7798	0.6095	0.4776	0.3751	0.2953	0.2330	0.1842	0.1460	0.1160

Table C.1. Present Value of $1.00

Present Value of an Annuity of $1 per Period at "*k*" rate for "*n*" Periods:

$$\frac{1 - \frac{1}{(1+k)^n}}{k}$$

Number of Periods	1%	2%	3%	4%	5%	6%	7%
1	0.9901	0.9804	0.9709	0.9615	0.9524	0.9434	0.9346
2	1.9704	1.9416	1.9135	1.8861	1.8594	1.8334	1.8080
3	2.9410	2.8839	2.8286	2.7751	2.7232	2.6730	2.6243
4	3.9020	3.8077	3.7171	3.6299	3.5460	3.4651	3.3872
5	4.8534	4.7135	4.5797	4.4518	4.3295	4.2124	4.1002
6	5.7955	5.6014	5.4172	5.2421	5.0757	4.9173	4.7665
7	6.7282	6.4720	6.2303	6.0021	5.7864	5.5824	5.3893
8	7.6517	7.3255	7.0197	6.7327	6.4632	6.2098	5.9713
9	8.5660	8.1622	7.7861	7.4353	7.1078	6.8017	6.5152
10	9.4713	8.9826	8.5302	8.1109	7.7217	7.3601	7.0236
11	10.3676	9.7868	9.2526	8.7605	8.3064	7.8869	7.4987
12	11.2551	10.5753	9.9540	9.3851	8.8633	8.3838	7.9427
13	12.1337	11.3484	10.6350	9.9856	9.3936	8.8527	8.3577
14	13.0037	12.1062	11.2961	10.5631	9.8986	9.2950	8.7455
15	13.8651	12.8493	11.9379	11.1184	10.3797	9.7122	9.1079
16	14.7179	13.5777	12.5611	11.6523	10.8378	10.1059	9.4466
17	15.5623	14.2919	13.1661	12.1657	11.2741	10.4773	9.7632
18	16.3983	14.9920	13.7535	12.6593	11.6896	10.8276	10.0591
19	17.2260	15.6785	14.3238	13.1339	12.0853	11.1581	10.3356
20	18.0456	16.3514	14.8775	13.5903	12.4622	11.4699	10.5940
21	18.8570	17.0112	15.4150	14.0292	12.8212	11.7641	10.8355
22	19.6604	17.6580	15.9369	14.4511	13.1630	12.0416	11.0612
23	20.4558	18.2922	16.4436	14.8568	13.4886	12.3034	11.2722
24	21.2434	18.9139	16.9355	15.2470	13.7986	12.5504	11.4693

Table C.2. Present Value of an Annuity of $1.00 per Period

Future Value of $1 at the End of "*n*" Periods, at "*k*" rate:

FV End = $(1 + k)^n$

Period	1%	2%	3%	4%	5%	6%	7%	8%	9%
1	1.0100	1.0200	1.0300	1.0400	1.0500	1.0600	1.0700	1.0800	1.0900
2	1.0201	1.0404	1.0609	1.0816	1.1025	1.1236	1.1449	1.1664	1.1881
3	1.0303	1.0612	1.0927	1.1249	1.1576	1.1910	1.2250	1.2597	1.2950
4	1.0406	1.0824	1.1255	1.1699	1.2155	1.2625	1.3108	1.3605	1.4116
5	1.0510	1.1041	1.1593	1.2167	1.2763	1.3382	1.4026	1.4693	1.5386
6	1.0615	1.1262	1.1941	1.2653	1.3401	1.4185	1.5007	1.5869	1.6771
7	1.0721	1.1487	1.2299	1.3159	1.4071	1.5036	1.6058	1.7138	1.8280
8	1.0829	1.1717	1.2668	1.3686	1.4775	1.5938	1.7182	1.8509	1.9926
9	1.0937	1.1951	1.3048	1.4233	1.5513	1.6895	1.8385	1.9990	2.1719
10	1.1046	1.2190	1.3439	1.4802	1.6289	1.7908	1.9672	2.1589	2.3674
11	1.1157	1.2434	1.3842	1.5395	1.7103	1.8983	2.1049	2.3316	2.5804
12	1.1268	1.2682	1.4258	1.6010	1.7959	2.0122	2.2522	2.5182	2.8127
13	1.1381	1.2936	1.4685	1.6651	1.8856	2.1329	2.4098	2.7196	3.0658
14	1.1495	1.3195	1.5126	1.7317	1.9799	2.2609	2.5785	2.9372	3.3417
15	1.1610	1.3459	1.5580	1.8009	2.0789	2.3966	2.7590	3.1722	3.6425
16	1.1726	1.3728	1.6047	1.8730	2.1829	2.5404	2.9522	3.4259	3.9703
17	1.1843	1.4002	1.6528	1.9479	2.2920	2.6928	3.1588	3.7000	4.3276
18	1.1961	1.4282	1.7024	2.0258	2.4066	2.8543	3.3799	3.9960	4.7171
19	1.2081	1.4568	1.7535	2.1068	2.5270	3.0256	3.6165	4.3157	5.1417
20	1.2202	1.4859	1.8061	2.1911	2.6533	3.2071	3.8697	4.6610	5.6044
21	1.2324	1.5157	1.8603	2.2788	2.7860	3.3996	4.1406	5.0338	6.1088
22	1.2447	1.5460	1.9161	2.3699	2.9253	3.6035	4.4304	5.4365	6.6586
23	1.2572	1.5769	1.9736	2.4647	3.0715	3.8197	4.7405	5.8715	7.2579
24	1.2697	1.6084	2.0328	2.5633	3.2251	4.0489	5.0724	6.3412	7.9111
25	1.2824	1.6406	2.0938	2.6658	3.3864	4.2919	5.4274	6.8485	8.6231

Table C.3. Future Value of $1.00 at the End of *n* Periods

Sum of an Annuity of $1 per Period for "$n$" periods:

$$\text{Sum of Annuity} = \frac{(1+k)^n - 1}{k}$$

Number of Periods	1%	2%	3%	4%	5%	6%	7%	8%	9%
1	1.000	1.000	1.000	1.000	1.000	1.000	1.000	1.000	1.000
2	2.010	2.020	2.030	2.040	2.050	2.060	2.070	2.080	2.090
3	3.030	3.060	3.091	3.122	3.153	3.184	3.215	3.246	3.278
4	4.060	4.122	4.184	4.246	4.310	4.375	4.440	4.506	4.573
5	5.101	5.204	5.309	5.416	5.526	5.637	5.751	5.867	5.985
6	6.152	6.308	6.468	6.633	6.802	6.975	7.153	7.336	7.523
7	7.214	7.434	7.662	7.898	8.142	8.394	8.654	8.923	9.200
8	8.286	8.583	8.892	9.214	9.549	9.897	10.260	10.637	11.028
9	9.369	9.755	10.159	10.583	11.027	11.491	11.978	12.488	13.021
10	10.462	10.950	11.464	12.006	12.578	13.181	13.816	14.487	15.193
11	11.567	12.169	12.808	13.486	14.207	14.972	15.784	16.645	17.560
12	12.683	13.412	14.192	15.026	15.917	16.870	17.888	18.977	20.141
13	13.809	14.680	15.618	16.627	17.713	18.882	20.141	21.495	22.953
14	14.947	15.974	17.086	18.292	19.599	21.015	22.550	24.215	26.019
15	16.097	17.293	18.599	20.024	21.579	23.276	25.129	27.152	29.361
16	17.258	18.639	20.157	21.825	23.657	25.673	27.888	30.324	33.003
17	18.430	20.012	21.762	23.698	25.840	28.213	30.840	33.750	36.974
18	19.615	21.412	23.414	25.645	28.132	30.906	33.999	37.450	41.301
19	20.811	22.841	25.117	27.671	30.539	33.760	37.379	41.446	46.018
20	22.019	24.297	26.870	29.778	33.066	36.786	40.995	45.762	51.160
21	23.239	25.783	28.676	31.969	35.719	39.993	44.865	50.423	56.765
22	24.472	27.299	30.537	34.248	38.505	43.392	49.006	55.457	62.873
23	25.716	28.845	32.453	36.618	41.430	46.996	53.436	60.893	69.532
24	26.973	30.422	34.426	39.083	44.502	50.816	58.177	66.765	76.790
25	28.243	32.030	36.459	41.646	47.727	54.865	63.249	73.106	84.701

Table C.4. Future Value of $1.00 per Period for *n* Periods (Sum of an Annuity of $1.00 per Period

Appendix D

Licensing of MTS Appraisers

Today, the licensing and certification of real estate appraisers is well established. In contrast, there are currently no federally mandated licensing requirements for the MTS appraiser or other personal property appraisers.[1]

Historically, developments in real estate appraisal have preceded those in other appraisal disciplines. To understand the direction in which licensing of MTS appraisers may be headed; it may be instructive to study the origin and history of the licensing of real estate appraisers.

What follows provides a historical perspective on the savings and loan crisis that occurred in the 1980s, the legislation and organizations that were formed as a result of the crisis, and the effects of these events on the appraisal profession. The current situation is described, including a brief description of the state certification and licensing agencies. The pros, cons, costs, and benefits of licensing are discussed. Additional references provided throughout the chapter will aid those who would like to obtain further information on the subject. This section does not address developments after early 1999.

Origins and History of Licensing of Real Estate Appraisers

The origins of existing licensing and certification of real estate appraisers can be traced to the savings and loan (S&L)

debacle in the United States, which culminated in massive failures of S&L institutions during the 1980s and early 1990s. Before the S&L crisis, the federal government did not impose licensing or certification requirements on appraisers, although a few states had licensing requirements. In most states, anyone could represent himself as an appraiser. Although codes of conduct and professional standards were maintained by various professional societies and organizations, appraisers and appraisals were essentially unregulated.

An important factor contributing to the failure of many S&Ls was the excessive loan-to-value ratios contained in their real estate loan portfolios. An excessive loan-to-value ratio occurs when the value of collateral held against a loan is less than the outstanding debt. Often, this condition is not realized until a borrower fails to pay on an outstanding loan. The lender then forecloses on the loan and liquidates the property to satisfy the debt. When the sale price of the property is insufficient to cover the amount of the outstanding debt, a loss is suffered. If the borrower cannot provide the difference in funds between the sale price and the loan amount, and the lender has no other recourse, the lender suffers the loss.

Fraud and incompetent management were two other factors precipitating the S&L failures. As the S&L crisis grew, the focus shifted to a common problem in many loan packages: the value of the collateral property was less than the outstanding debt. There are two primary reasons for this situation: (1) overstated property values, which resulted in lenders overadvancing against the collateral pledged on the loan; and (2) improper property liquidation methods, which brought less than market value at the liquidation sale of the property. Overstated property values were in part based upon fraudulent or overly optimistic appraisals supplied to overzealous lenders, who in turn granted excessive loans. An improper liquidation showed that the collateral was sold at less than market value through poor sales techniques, incompetence, or fraud.

Appraisers naturally argued that poor collateral sales were the problem, and property liquidators and auctioneers blamed faulty appraisals. Ultimately, both conditions were shown to exist.

Federal Actions and the Appraisal Subcommittee

The massive S&L failures, which involved the lending of funds insured by the Federal Deposit Insurance Corporation (FDIC), necessitated federal government involvement in the crisis. As insured losses soared, the federal government attempted to shore up the collapsing S&L industry.

In August 1989, Congress passed the Financial Institutions Reform Recovery and Enforcement Act of 1989 (FIRREA). This act established the Appraisal Subcommittee (ASC) of the Federal Financial Institutions Examination Council.

The ASC has six members, each designated respectively by the heads of the Office of the Controller of the Currency, the Board of Governors of the Federal Reserve System, the FDIC, the Office of Thrift Supervision and the National Credit Union Administration, collectively; the federal financial institutions regulatory agencies; and the U.S. Department of Housing and Urban Development.

Section 1103 of Title XI sets out the ASC's general responsibilities:

- Monitor the requirements established by the States, territories, and the District of Columbia ("States") and their appraiser regulatory agencies ("state agencies") for the certification and licensing of appraisers. The ASC reviews each State's compliance with the requirements of Title XI and is authorized by Title XI to take action against non-complying States;

- Monitor the requirements established by the Agencies regarding appraisal standards for federally related transactions and determinations of which federally related transactions will require the services of State licensed or State certified appraisers;

- Maintain a national registry of State licensed and certified appraisers (“Registry”) who may perform appraisals in connection with federally related transactions;
- Monitor and review the practices, procedures, activities and organizational structure of the Appraisal Foundation (“Foundation”); and
- Transmit an annual report to Congress regarding the activities of the ASC during the preceding year.[2]

Appraisal Subcommittee Study

Section 1122(e) of Title XI of FIRREA mandated that the ASC conduct a study to determine the feasibility and desirability of extending the provisions of Title XI to the function of personal property appraisal and personal property appraisers in connection with federal financial and public policy interests. The ASC conducted such a study and published a report of its conclusions in March 1991 under the title *Personal Property Appraisal Study.* The report was prepared from surveys. The majority of public comments and responses came from institutional appraisal users. There were 666 respondents, including 415 state banks, 111 national banks, 13 credit unions, 10 savings institutions or savings banks, and 33 financial institutions’ trade associations. The remaining 84 respondents were holding companies, appraisal firms, individuals, trade organizations, state and federal agencies, and other individuals or miscellaneous companies.

The report was divided into three sections: public comments, feasibility, and desirability. The following discussion recapitulates the report contents and conclusions. The results provide a broad spectrum of viewpoints related to the licensing of personal property appraisers and a look at how the decision was made to *not* recommend federally mandated licensing for personal property appraisers.

In the public comments section, one of the first things included is a definition of personal property: “identifiable portable and tangible objects which are considered by the general public as being 'personal,' e.g., furnishings, artwork, antiques, gems and jewelry, collectibles, machinery and equipment: all property that is not classified as real estate."

The ASC questions for public comment focused specifically on what types of personal property collateral might be responsible for the greatest losses, or conversely, for having the best record. The subcommittee also attempted to determine whether there was a direct link between losses and faulty evaluation of collateral. Most of the responses received were nonspecific as to property types and anecdotal in nature. Automobiles were mentioned as a large loss generator. Automobiles, however, represented a very large percentage of the total outstanding loans and these conditions basically balanced. There were specific comments regarding agriculture-related products and equipment losses. This was consistent with a large response from Midwestern lenders who had concentrations of agricultural related equipment in their portfolios. The public comments section of the subcommittee report states:

The responses almost unanimously indicate that losses have not been sustained as a result of faulty evaluation of personal property collateral. While losses were suffered on personal property lending, the following reasons were given for those losses:

- changes in prevailing market conditions and resultant inability of borrower to make interest and principal payments,
- deterioration in the value of the underlying collateral at a rate faster than the loan amortizes,
- drastically lower liquidation value than the original valuation,
- failure to perfect a security interest in the collateral,
- disappearing collateral or substitution of the original collateral, and

- poor maintenance of collateral resulting in accelerated depreciation.[3]

Also detailed in the public comments section was the respondents' general perception that, to a large extent, Title XI requirements were not applicable to personal property appraisals. For personal property loans, secondary sources of reliance and repayment have to be taken into account. Consideration has to be given to the other basic elements of credit (the "five Cs")—capacity, credit history, character, and common sense, along with collateral. Many fundamental differences between real estate and personal property loans were identified, namely the great diversity of personal property (for example, gems, jewelry, and machinery and equipment). Each of these areas is unique and would require special consideration for licensing purposes. The many different premises of value for personal property were mentioned. It was also pointed out that loan-to-value ratios for personal property loans were usually lower than those for real estate.

The public comments section also addressed personal property appraisal standards. This section mentioned standards 7 and 8 of USPAP, which deal with personal property appraisers and appraisals. The study concluded that the field is too diverse to develop one set of meaningful uniform standards. Another opinion, endorsed primarily by appraisers and appraisal associations, supported the development of one set of uniform standards, generic in nature. The study pointed out that Title XI specifically defines the licensing procedures and qualification criteria for real estate appraisers.

The debate continues about whether licenses should be specific to property type or generic for all personal property. The consensus of those who supported generic standards is a single licensing process. Ultimately, most respondents believed that the time and cost required to develop, implement, and administer a personal property certification and licensing program would outweigh the benefits to the public, considering there have not been sustained lending losses because of faulty valuations.

Questions relating to the regulation and supervision of personal property appraisers drew inconsistent responses. The basic questions were

- Who should develop standards, qualifications, and testing criteria?
- Who should have supervisory enforcement of the standards if developed?

The study showed a preference for personal property appraisal organizations and financial institutions taking control instead of regulatory agencies or legislative bodies. It was suggested that the appraisal organizations assume responsibility for qualification and testing and that their registry fees be affordable, yet high enough to cover supervision costs. Most commentary did not support a primary role for federal agencies in the enforcement of standards, but a majority did support federal monitoring.

It was reported that 643 of the 666 respondents opposed extending the real estate appraisal regulatory framework to personal property appraisal. Seventeen supported it and six took no stance. It is interesting to note that the opposition to the licensing of personal property appraisers came mainly from appraisal users. Supporters of licensing were mainly those within the appraisal community.

The second section of the study dealt with the feasibility of certification and licensing for personal property appraisers. The ASC determined that *it is feasible* to certify and license personal property appraisers in a framework similar to Title XI. However, comments were made that were similar to those in the public comment section. Two comments in the ASC study under the feasibility section are (1) personal property is too diverse in nature to allow for one set of meaningful uniform standards; and (2) it is difficult to obtain qualified appraisals in small and rural communities because of the limited number of personal property appraisers.

The ASC reasoned that implementation is feasible only if the standards are very generic in nature. The subcommittee did not believe that categories (or even groupings of categories) were feasible. The basic conclusions were:

- Generally, the valuation of collateral for loan purposes is done in specific steps. In the broadest sense, the property is identified, the purpose of the appraisal understood, and the value estimated.

- Although the development of generic standards is feasible, implementation of licensing or certification requirements would be far more difficult. A wide range of people who are not designated appraisers presently provide valuation services. For instance, bank employees regularly use automobile price guides to determine the collateral value of automobiles. Any attempt to license or certify such individuals is feasible only over an extended time period.

The ASC concluded that the extension of the real estate appraisal framework for licensing or certification was feasible but stressed that the standards and competency requirements should be generic. They estimated that it would take three to five years to implement even minimal personal property appraisal standards and competency criteria.

The results from the desirability portion of the study led the ASC to conclude that certification and licensing of personal property appraisers is *neither desirable nor necessary.* The primary reason for this finding was that the committee had been "unable to document any recurring patterns of abuse or evidence that significant systemic losses have been sustained by insured financial institutions, the Resolution Trust Corporation, the FDIC, or the National Credit Union Share Insurance Fund as a result of faulty valuation of personal property."

The comment section of the study focused on two issues:

- No evidence showed that losses have resulted from loans where personal property served as collateral.

- No reason exists to believe that regulation of personal property appraisals or appraisers would have prevented any losses that have occurred.

These comments indicate the ASC recognized that there are inherent differences in the nature of personal and real property and that lending practices differ between them. Most lenders have historically viewed personal property loans to be more risky than real estate transactions. Real estate is less likely to suffer loss in value through use and less likely to suffer uninsured destruction. Personal property lenders typically require larger down payments and offer shorter-term loans.

The automotive financial institutions are the largest group of personal property lenders. They are followed by equipment lessors, who lend a large percentage of the money used in personal property financing. These kinds of institutions typically have internal staff evaluate exposure in the loan and lease portfolios and often do not retain outside appraisers.

The federal government's position on the certification and licensing of personal property appraisers remains unchanged from the time of the study. Nevertheless, some states and some professional appraisal organizations continue to debate the subject. Each of the states has regulating bodies created in response to Title XI. At a time when many government jobs are being eliminated and spending is being cut, it is possible that a new source of revenue from professional fees would receive special attention from state government bureaucracies.

The Appraisal Foundation

The Appraisal Foundation is another entity that was created as a result of the S&L crisis. In the mid-1980s, as the S&L crisis grew, it became clear to the real estate appraisal community that government regulation of some sort was forthcoming. Concerns over the practicality of impending regulations led to the formation of an ad hoc committee of appraisal practitioners. The group was primarily composed of real estate appraisers because the losses suffered in the S&L crisis were related to real estate development and loans. The goal of the ad hoc committee was to provide input to government regulators from a united appraisal voice. The intent of the committee was to aid in the development of sensible requirements for the appraisal industry. This ad hoc committee grew to become the Appraisal Foundation, which was incorporated as a not-for-profit corporation in Illinois in November 1987.

By the time the Appraisal Foundation was incorporated, the ad hoc committee had already developed a set of uniform standards for the appraisal profession. These same standards were copyrighted in 1987 by the Appraisal Foundation as the *Uniform Standards of Professional Appraisal Practice* (USPAP). Students of the appraisal profession must be aware of USPAP because these standards have been adopted by virtually all professional appraisal organizations in the United States, by state and federal agencies, and by some foreign governments and organizations. The Appraisal Foundation continues to work closely with federal financial institutions, regulatory agencies, and the ASC.

The Appraisal Foundation is now the primary developer of USPAP. In addition to appraisers, it serves the ASC, other governmental bodies, users of appraisal services, and the public at large. The Foundation comprises representatives from various professional appraisal organizations. It has significant influence on regulation of the appraisal community. This is especially true with respect to the licensing, certification standards, and qualifications of appraisers.

In 1989, the Appraisal Foundation established two groups to oversee the development and continuing administration of standards and professional criteria for appraisals and appraisers:

- Appraisal Standards Board (ASB)
- Appraiser Qualifications Board (AQB)

The ASB was given the charter to develop, publish, interpret, and amend appraisal standards. The ASB adopted USPAP and continues to develop and administer it as an evolving document.

The AQB was formed to develop, publish, interpret, and amend qualification criteria for professional appraisers. In March 1991, it adopted criteria for real estate appraisers. The criteria are included in the publication *Interpretations and Clarifications of the Criteria.*

In 1992, the Office of Management and Budget (OMB) issued Bulletin 92-06 regarding the use of real estate appraisals by the agencies under their jurisdiction. Among other things, the bulletin adopted the AQB criteria as the level of training its agencies should provide to its appraisal employees. In 1994, the AQB announced that it had decided to pursue the development of personal property appraiser criteria. In 1996, the AQB announced the same intention for business valuation. Many people believe these announcements signal a move in the direction of licensing for appraisers of property other than real estate.

Federal License Requirements

With the enactment of FIRREA in 1989, the federal government mandated that all states develop appraiser certification and licensing bodies. Title XI of FIRREA established the ASC as the federal agency to monitor and supervise these bodies. The Appraisal Foundation and its boards have been recognized as authorities by federal and state regulators. The USPAP and AQB criteria have been adopted as minimum requirements. Federal requirements for the licensing and certification of appraisers currently do not exist. Real estate appraisers are licensed and regulated only at the state level. The National Registry of State Certified and Licensed Appraisers is maintained by the ASC in compliance with Title XI of FIRREA.

Federal and state regulation of the real estate appraisal community was developed between 1987 and 1995. In its annual report to Congress in January 1994, the ASC reported that "1993 was the first year that Title XI was fully in place and related federal and state regulatory programs were in full operation." In retrospect, much has occurred in a short period. Increased regulation and additional requirements are likely to occur now that the structure for regulation is in place in the form of state agencies, the Appraisal Foundation, and the ASC.

State Certification and Licensing Regulatory Agencies

Each state has a designated agency to administer the certification and licensing of real estate appraisers. Each state has established minimum requirements in terms of education, experience, and testing in accordance with Title XI of FIRREA. These minimums equal or exceed the criteria established by AQB and standards established by the ASB of the Appraisal Foundation. As part of its certification and licensing program, each state is required to provide the ASC with a list of all of its certified and licensed appraisers. The list and required fees are collected and supplied to the National Registry of Certified Appraisers.

The ASC opined that Title XI "intends that states supervise all of the activities and practices of persons who are certified or licensed to perform real estate appraisals in connection with *all* real estate appraisals involving real estate related financial transactions, and not just federally related transactions." This statement, published in the January 1994 annual report to Congress, further states that state agencies investigate inappropriate behavior on the part of licensed appraisers and take disciplinary actions.

Many states have already gone beyond the basic requirements of FIRREA and Title XI with respect to certification and licensing of appraisers. The information in table D.1 was issued by the ASC in 1993 and was published by the Appraisal Foundation in its *Appraiser Update* in September 1993.

Alabama	M/FRT	Nebraska	M
Alaska	M/FRT	Nevada	M
Arizona	M	New Hampshire	V
Arkansas	M/FRT	New Jersey	M/FRT
California	M/FRT	New Mexico	M
Colorado	M	New York	V
Connecticut	M/FRT	North Carolina	V
Delaware	M	North Dakota	V
District of Columbia	M	Ohio	V
Florida	M/FRT	Oklahoma	V
Georgia	M	Oregon	M
Hawaii	M/FRT	Pennsylvania	M/FRT
Idaho	V	Rhode Island	V
Illinois	V	South Carolina	M
Indiana	V	South Dakota	V
Iowa	V	Tennessee	M
Kansas	M/FRT	Texas	M/FRT
Kentucky	V	Utah	M
Louisiana	V	Vermont	V
Maine	M/FRT	Virginia	M
Maryland	M/FRT	Washington	V
Massachusetts	M/FRT	West Virginia	M
Michigan	M	Wisconsin	V
Minnesota	M	Wyoming	V
Mississippi	M	Puerto Rico	M
Missouri	V	Virgin Islands	M
Montana	V		

M	Mandatory	All appraisals must be done by a licensed or certified appraiser.
M/FRT	Mandatory for Federally Related Transactions	All appraisals must be done by a licensed or certified appraiser.
V	Voluntary	It is left to the discretion of lenders as to whether an appraiser must be licensed or certified to do appraisal work. Lenders, however, must abide by Title XI.

Table D.1 State Licensing Requirements

Each state has its own entity for the licensing of real estate appraisers. Each state charges a fee for appraiser certification or licensing. To conduct an appraisal in a federally related real estate transaction in any state, the appraiser must be licensed in that state or be recognized for temporary practice by that state. Title XI requires state agencies to recognize, on a temporary basis, the certification or licensing of an appraiser from another state, provided that (1) the property to be appraised is part of a federally related transaction; (2) the appraiser's business is of a temporary nature; and (3) the appraiser registers with the state appraiser regulatory agency in the state of temporary practice. A nominal per-assignment fee can be charged by the state granting the temporary practice permit.

The entire certification and licensing process has become a significant expense for real estate appraisers. The costs of regulation have been borne by the profession, a result inherent in the enactment of FIRREA. The Department of the Treasury advanced the initial federal costs for the ASC with the expectation of being reimbursed after fees were collected. The fees for an appraiser who practices in more than one state can be expensive. Some appraisers believe these fees are excessive and prohibit them from having a multistate practice.

Reciprocity of licenses between states is not the same as a temporary practice permit. Reciprocal arrangements occur as one state accepts part or all of the certifying and licensing requirements of another state. Not all states share reciprocal rights. Real estate appraisers who practice in more than one state must be sure that they have fulfilled each state's requirements.

There are different forms of reciprocal arrangement between states. One arrangement is similar to reciprocal driver's privileges between states. Under this type of arrangement, a certified and licensed real estate appraiser from another state must simply possess a valid certificate and be competent to perform the assignment under USPAP. No testing or additional fees are charged. Another type is a formal reciprocal agreement between states. A certified real estate appraiser in good standing in one state applies and is granted certification or licensing in a second state upon review of the appraiser's existing license and credentials and receipt of licensing fees. The second state is essentially accepting the first state's education requirements and examination process instead of its own; however, it charges a fee to the appraiser. The third type of reciprocity is an agreement in which, in addition to the fee, some additional level of education, experience, or testing is required.

All of the reciprocity agreements permit continued appraisal activity, usually for one year. The requirement for real estate certification and licensing in each state is difficult for appraisers who continually practice on a multistate basis. An appraiser who has a client with facilities in more than one state must be certified in each of those states. This is also the case for many appraisers who provide allocation of purchase price appraisals of acquired business entities in other states.

The ASC stated in its 1993 annual report that it endorses reciprocity between states and urges states to establish permanent reciprocity arrangements to address the needs of appraisers with multistate, nontemporary practices. A reciprocal agreement similar to the "drivers license" type would be preferred by most appraisers. This kind of arrangement would be ideal for personal property appraisers, especially in light of the past ASC study which indicated little need for personal property appraiser licensing.

Current Situation and Opinions[4]

In February 1995, the AQB held a public hearing regarding criteria for personal property appraisers. At that hearing Richard Kaufman, then the international president of the American Society of Appraisers, made formal remarks to the Board. An article containing those remarks was published in ASA's March 1995 issue of *Newsline*. Also attending the hearing were the chairperson of the Education Committee of the International Society of Appraisers, the executive director of the Appraisal Association of America, and the director of the Association of Machinery and Equipment Appraisers. Formal presentations were given by representatives from the U.S. Marshal's Service and the Equipment Leasing Association. All of the panelists spoke in general support of establishing qualification criteria for personal property appraisers but advised that a careful and deliberate process be followed. The following excerpts are from the *Newsline* article and Mr. Kaufman's remarks to the AQB. The comments address some of the problems with real estate appraisal licensing in 1995 and also provide insight on prevailing thoughts on the development of standards, criteria, and regulations for personal property and other non-real-property appraisers.

- The ASA does not see a workable personal property qualification environment based on the real property requirements of Title XI.

- Real property appraisers did not deserve the amount of regulations that has evolved from Title XI.

- The ASA would be in favor of a national registry of qualified personal property appraisers, similar to the Internal Revenue Service Supervision of Enrolled Agents and Actuaries or the Securities and Exchange Commission Registration of Investment Advisers. A national registry of qualified personal property appraisers would facilitate freedom of temporary practice in any jurisdiction, a privilege that the real property appraisers do not currently enjoy.

- The ASA supports the development of guidelines for personal property appraisal users. Such guidelines should be established on a mutually and readily agreeable basis between both the appraiser and the user. If both groups cannot accept a guideline, then it should not be established.

- There is no pressing national policy or public issue like the S&L crisis that would require personal property appraisers to be similarly regulated.

- Initial personal property appraiser qualification guidelines might address the following issues:
  - Do they conduct their assignments in accordance with generally available and relevant standards?
  - What general and specific education have they had and how does it enhance their appraisal abilities?
  - What specific experience do they have with the subject of the appraisal, the purpose of the appraisal ,and the function of the appraisal?
  - What references can they provide to verify their expertise? References might include previous clients, recognition by courts or other tribunals, or appropriate professional designations by scientific, artistic, and other professional societies.
  - Are they properly registered as businesses in the communities or states in which they serve?
  - Do they acknowledge the limitations of the appraisal discipline—real property, personal property, or business appraisal—in which they choose to practice? Such distinctions have already been recognized by the Appraisal Foundation and the ASB.

Looking Ahead

No easy answers are available for questions regarding the status of certification and licensing for appraisers of disciplines other than real estate. *No one* knows what the future may hold. The process is ongoing. Interested appraisers and appraisal students should pay close attention to developments within the ASC and the Appraisal Foundation because it is these two entities that are recognized by regulators of the appraisal industry.

Notes

[1] As of early 1999, when this was written, the licensing of real property appraisers had resulted in some ad hoc regulation of some MTS appraisers in at least some states. In some states, large items of machinery and equipment are defined as real property; some judicial and administrative bodies in these states require an appraiser testifying as an expert witness (e.g., in property tax appeals) to possess a real property appraisal license.

[2] Appraisal Subcommittee, Annual Report 1999 (Washington, D.C.: Appraisal Subcommittee, 1999), p. 1.

[3] Ibid.

[4] This section was written in 1998 and updated in early 1999.

Appendix E

Published Prices and Sources of Information

Aero Tracer
P.O. Box 9059, St. Petersburg, FL 34618
800-548-8889

Aircraft Blue Book
Aircraft Blue Book Corporation
Box 12901, Overland Park, KS 66212
800-654-6776

Aircraft Blue Book – Price Digest
Aircraft Blue Book Corporation
Intertec Publishing Corporation
Will Rogers World Airport
P.O. Box 59977, Oklahoma City, OK 73159

Aircraft Value Newsletter
Phillips Business Information, Inc.
1925 North Lynn Street, Suite 1000, Arlington, VA 22209
803-522-8333

Airliner Price Guide of Commercial-Regional and Commercial Aircraft
Airliner Price Guide
1105 Sovereign Row, Oklahoma City, OK 73108
405-942-8225

American Hotel Registry
2775 Shermer Road, Northbrook, IL 60062-7798
800-323-5686

Appraisal Principles and Procedures
Babcock, Henry A.,
Published by Richard D. Irwin
Homeward, IL, 1968

Assessor's Handbook General Appraisal Manual
Assessment Standards Division, Property Tax Department
Published by California State Board of Equalization, CA, 3/95

Atlantic Flyer
P.O. Box 39099, Tacoma, WA
20-) 471-9888
Aviators' Hotline

Hotline Publications
P.O. Box 958, 1003 Central Avenue, Fort Dodge, IA 50501
800-247-2000

Avionics News
Published by Aircraft Electronics Association
13700 East 42nd Terrace, Suite 102, Independence, MO 64055
816-373-6565

Black Book
National Auto Research
Div. Of Hearst Business Media Corporation
P.O. Box 758, Gainesville, GA 30503
404-532-4111

Blue Book – Auto
Kelly Market Report
P.O. Box 19691, Irvine, Costa Mesa, CA 92718
800-258-3266

Blue Book of Current Market Prices of Used Heavy Construction Equipment
Forke Brothers,
830 NBC Center, Lincoln, NE 68508

Boats and Harbors
P.O. Drawer 647, Crossville, TN 38557
616-484-8708

The Book (Used Machinery Pricing Guide, Auction)
L&M Publications, Inc.
318 Oak St. N.W.
P.O. Box 3273, Gainesville, GA 30503
800-262-8005, 404-532-6510

Buc Research (Used Boat Price Guides)
Buc International Corporation
1314 N.E. 17th Court
Fort Lauderdale, FL 33305
800-327-6929

Building and Construction Costs
Stevens Valuation Quarterly,
Published by Marshall and Swift Publication Company,
911 Wilshire, 16th Floor, Los Angeles, CA 90017-3409
213-683-9000

Building Construction Cost Data
R.S. Means & Co., Inc.
100 Construction Plaza, Kingston, MA 02364
617-585-7898

Chemical Marketing Reporter
Schnell Publishing Company

Computer Hardware & Software
Data Services
P.O. Box 5845, Cherry Hill, NJ 08034

Computer Hot Line
15400 Knoll Trail, Dallas, TX 75248
800-999-5131, 214-233-5131

Computer Merchants
22 Saw Mill River Road, Hawthorne, NY
914-592-1060

Computer Price Guide
The Blue Book of Used IBM Computer Prices
200 Brady Avenue, Hawthorne, NY 10532
914-769-2686

Construction Equipment Advertiser
P.O. Box 7767, Long Beach, CA 90807
310-595-5731

Contractors Hotline
Heartland Communications Group, Inc.
Box 1052, 1003 Central Avenue
Fort Dodge, IA 50501
800-247-2000

Energy Cost Reference Book
AACE
308 Monongahela Building, Morgantown, WV 26505

Equipment Directory of Audio Visual Computer and Video Products
International Communications Industries Association
3150 Spring Street, Fairfax, VA 22031

Equipment Journal
150 Lakeshore Road W., #36, Mississauga Ontario, L5H 3R2 Canada
905-274-4883

Equipment Today
Johnson Hill Press, P.O. Box 803, Fort Atkinson, WI 53538-0803, 414-563-1745

Equipment World
Randall Publishing, Inc., 3200 Rice Mine Road, NE, Tuscaloosa, AL 35406
800-633-5953

Equipment World
P.O. Box 2029, Tuscaloosa, AL 35403
205-349-2990, 205-750-8070

Estimators Equipment Installation Man Hour Manual
Gulf Publishing Co., Houston, TX.

Farm Equipment Guide – Hotline
Heartland Communications Group, Inc.
Box 1052, 1003 Central Avenue
Fort Dodge, IA 50501
800-247-2000

Green Guide Auction Reports
Dataquest, Inc.
1290 Ridder park Drive, San Jose, CA 95131-2398
800-669-3282

Green Guide for Lift Trucks
Dataquest, Inc.
1290 Ridder Park Drive, San Jose, CA 95131
408-437-8001

Industrial Machinery Digest
205-988-9708

Industrial Machinery Trader
Heartland Communications Group, Inc.
Box 1052, 1003 Central Avenue
Fort Dodge, IA 50501
800-247-2000

Industrial Market Place
7842 North Lincoln Ave., Skokie, IL 60077
800-676-1900, 708-676-1900

International Aircraft Price Guide
Published by Aviation Information Services, Ltd., Cardinal Point, Newhall Road, Heathrow Airport (London), Hounslow TW6 2AS
441-897-0615/0300

Jet Aircraft Values and Turbo Prop Values
Published by Avitas Bluebooks, 1893 Preston White Drive, Suite 210, Reston, VA 22090
703-476-2300

Kelly Blue Book
P.O. Box 19691, Irvine, CA 92713
800-258-3266

The Last Bid
The Last Bid International Equipment Exchange
6689 Peachtree Industrial Boulevard, Suite P,
Norcross, GA 30092
404-447-8473

Locator
Machinery Information Systems
1110 Spring Street, Silver Spring, MD 20910-4056

Machinery and Equipment Pricing Guide
Published by Construction Publishing Company, Inc., New York, NY
Machinery Journal
P.O. Box 7767, Long Beach, CA 90807
310-595-5731, fax 310-424-1563

Machinery Journal
310-595-5731

Machinery Trader
P.O. Box 85010, Lincoln, NE 68501-9967
800-334-7445

Marine Equipment Catalog
Maritime Activity Reports, Inc., 118 East 25th Street, New York, NY 10010
212-477-6700

Marshall Valuation Service
Marshall & Swift
911 Wilshire, 16th floor
Los Angeles, CA 90017-3409
213-683-9000

Mine and Quarry Trader
P.O. Box 603, Indianapolis, IN 46206
800-827-7468, 317-297-5500

My Little Salesman
P.O. Box 70208, Eugene, OR 97401
800-352-0642

N.A.D.A. Retail Aircraft Appraisal Guide
Published by N.A.D.A. Appraisal Guides, P.O. Box 7800, Costa Mesa, CA 92628
800-966-6232

National Equipment Digest
Vulcan Publications, Inc., P.O. Box 55886, Birmingham, AL 35255
800-877-9748

Nomda's Blue Book Industry Guide to Market Prices & Manufacturers' Trade-In Schedules
National Office Machine Dealers Association
P.O. Box 707, Wooddale, IL 60191

Official Guide - Tractors and Farm Equipment
Southwest Hardware and Implement Association
4629 Mark IV Parkway, Fort Worth, TX 76106

The Official Helicopter Bluebook
Helivalues, Inc., P.O. Box 876, Lincolnshire, IL 60069-0876
708-634-3877

Orion Research Corp
14555 N. Scottsdale Road, Suite 330, Scottsdale, AZ 85254
602-951-1114

Plastic Hotline
Heartland Communications Group, Inc.
Box 1052, 1003 Central Avenue
Fort Dodge, IA 50501
800-247-2000

Prices of New Machine Tools
Office of Price Administration, Published by Office of Price Administration, U.S.

Processor (Computer Equipment)
P.O. Box 85010, Lincoln, NE 68501-9967
800-334-7445

Process Plant Construction Estimating Standards
Richardson Engineering Services, Inc.
4 Volumes
P.O. Box 9108, Mesa, AZ 85214-9103
602-497-2062

Rock & Dirt
410 West Fourth Street, Crossville, TN 38557
800-251-6776, 615-484-5139

Rule of Thumb Cost Estimating Building MechanicalSystems
Konkel, James H., Published by McGraw-Hill Publications, New York, NY, 1987

Surplus Record
20 North Wacker Drive, Chicago, IL 60606-9757
800-622-5449

Top Bid
P.O. Box 2029, Tuscaloosa, AL 35403
800-633-5953

Top Bid
Randall Publishing Co., 3200 Rice Mine Road, N.E., P.O. Box 2029, Tuscaloosa, AL 35403
800-633-5953, 205-349-2990, fax 205-249-3765

Truck Blue Book
Forke Brothers
P.O. Box 21960, Lincoln, NE 68542

The Truck Blue Book
National Market Reports, Inc.
300 West Adams Street, Chicago, IL 60606
312-726-2802

Truck Gazette
P.O. Box 7767, Long Beach, CA 90807
310-595-5731, fax 310-424-1563

Truck Paper
P.O. Box 85010, Lincoln, NE 68501-9967
800-334-7445

Used Equipment Directory
P.O. Box 823, Hasbrouk Heights, NJ 07604

Appendix F

Recommended Readings

Many are available through public libraries or at a reasonable cost from the publishers.

Accounting
Ronald Press

AG Consultant
37733 Euclid Ave., Willoughby, OH 44094-5925
216-942-2000, fax 216-942-0662

Agri Finance
6201 Howard Street, Niles, IL 60714
708-647-1200, fax 708-647-7055

Agri Marketing
6201 Howard Street, Niles, IL 60714-3435
708-647-1200, fax 708-647-7055

Airline Business
Reed Business Publishing, Cahners Plaza, 1350 East Touhy Avenue, Suite 235 W., Des Plaines, IL 60018

Airline Financial News
Phillips Business Information Inc., 1925 North Lynn Street, Suite 1000, Arlington, VA 22209
703-522-8333

American Sailor
P.O. Box 209, Newport, RI 02840
401-849-5200, fax 401-849-5208

Appraisal Principles and Procedures
Babcock, Henry A., Published by Richard D. Irwin, Homeward, IL, 1968

Appraising Machinery and Equipment
Sponsored by the American Society of Appraisers
McGraw-Hill Book Company

An Approach To A Method of Evaluating Chemical Plants and Refineries
Redding, E.D., A.S.A. Appraisal and Valuation Manual, Published by American Society of Appraisers, 1959

Assessor's Handbook General Appraisal Manual
Assessment Standards Division, Property Tax Department. Published by the California State Board of Equalization, California, 3/75

Audels-Machinists and Tool Makers Handy Book
Frank Graham, New York, NY

Audels-Power Plant Engineers Guide
Frank Graham, New York, NY

Avmark Newsletter
Avmark, Inc., Lewis & Clark Trail, Lolo, MT 59847
406-273-2580

Basic Prices of Machine Tools
Office of Price Stabilization, Published by U.S. Government Printing Office, February 1953

The Big Book of Metalworking Machinery
Sylmar California

Boat & Motor Dealer
3949 Oakton, Skokie, IL 60075-3460
708-982-1810, fax 708-675-7402

The Boat Broker
8365 Old Dairy Road, Juneau, AK 99801-8041
907-789-5581, fax 907-789-0987

Boating
1633 Broadway, 43rd Floor, New York, NY 10019
212-767-6000, fax 212-767-5618

Boating Business
447 Speers Road, #4, Oakville, Ontario L6K, 3S7 Canada
905-842-6591, fax 905-842-6842

Boating Industry
5 Penn Plaza, 13th Floor, New York, NY 10001-1810
212-613-9700, fax 212-613-9749

Boating World
2100 Powers Ferry Road, Atlanta, GA 30339-5014
404-955-5656, fax 404-952-0669

Business Air Today
Business Air Today, 1003 Central Avenue, Fort Dodge, IA 50501
800-892-0267

Business Aviation Weekly
McGraw-Hill Aviation Weekly Newsletters, Box 666, Highstown, NY 08520
800-551-2014

The Business Farmer
1617 Avenue A. Scottsbluff, NE 69361-3163
308-635-2045, fax 308-635-2348

Business Information Sources
Lorna M. Daniels, University of California Press, Berkeley and Los Angeles, California, 1985

California Expert Witness Guide
Raoul D. Kennedy, Esq., California Continuing Education of the Bar, Berkeley, California

The Car & Locomotive Cyclopedia
Published by Simmons Boardman Books
402-248-0300

Commuter World
Shephard Press, ltd., 111 High Street, Burnham, Bucks, SL1-7J2 England

The Contemporary Diesel Spotter Guide
Diesel Locomotive Rosters
Diesel Spotters' Guide
Kalmback Publishing
800-533-8644

Controller
Peed Corporation, P.O. Box 85310, Lincoln, NE 68501
800-247-4890

Controlling Software Projects
Yourdon Press

Cooperative Farmer
6606 W. Broad Street, Richmond, VA 23230-1717
804-281-1317, fax 804-281-1141

Cost Estimating Manual for Pipelines and Marine Structures
Gulf Publishing, 1977

Creating Computer Software User Guides
McGraw-Hill Book Company, New York, NY

Dictionary of Scientific and Technological Terms
McGraw-Hill Book Company
Highstown, NJ 08520

Directions - Cessna Citation Magazine
Cessna Aircraft Company, P.O. Box 7706, Wichita, KS

Director of Industry Data Sources
Ballinger Publishing Company
Division of Harper & Row, Cambridge, MA

Dodge Construction and Pricing Manual
McGraw-Hill Information System Company, New York, NY

Electronic Design's Gold Book
3-Volume Set, Gold Book Company, Philadelphia, PA 19101

Engineering Economy
Gerald W. Smith, The Iowa State Press, Ames, IA

Engineering Valuation and Depreciation
Marston Winfrey & Hempstead, Ames, Iowa, The Iowa University Press

Environmental Impact Studies; Their Need and Components
Allman, Storm A., Published by Marshall and Stevens, Inc., August 1973

Equipment Leasing Today
1300 N. 17th Street #1010, Arlington, VA 22209-3875
703-527-8655, fax 703-527-2649

Equipment Today
1233 Janesville Ave. Fort Atkinson, WI 53538-2738
404-563-6388, fax 414-563-1699

Equipment Financing Journal
716-885-0444

European Rubber Journal
Crain, 1725 Merriam Road, Akron, OH 44313, Robert S. Simmons, V.P. Publications Director
216-836-9180, fax 216-836-2322

Field Manual of the A.A.R. Interchange Rules
Association of American Railroads, 50 F Street N.W., Washington, DC 20001-1564

Flying
Hachete Magazines, Inc., 1633 Broadway, New York, NY 10019
203-622-2725

Food Processors Guide
Food Processing Machinery and Supplies Association
1828 L. Lane, N.W., Washington, DC 20036

Gas Daily
1616 N. Fort Myer Drive, #1000, Arlington, VA 22209-2306
703-528-1244, fax 703-528-1253

Industrial Market Place
7842 N. Lincoln Ave., Shokie, IL 60077
800-323-1818, fax 708-676-0063

International Marine Dictionary
Rene de Kerchone, Van Norstrand Reinhold Company, 1961, New York, NY

Life Cycle Cost Data
Alphonse J. Dellisola and Stephen J. Kirk, McGraw-Hill Book Company, New York, NY

Machinery Outlook
1110 Lake Cook Road, #295, Buffalo Grove, IL 60089-1965
708-215-2999, fax 708-215-0455

Maritime Reporter and Engineering News
Maritime Activity Reports, Inc., 118 East 25th Street, New York, NY 10010
212-477-6700

Mining Magazine
Mining Journal, Ltd., 60 Worship Street, London, EC 2A 2HD, U.K.
44-632-864-333

Modern Plastics Magazine
McGraw-Hill, 1221 Sixth Avenue, New York, NY 10020, Keith R. Kreisher, Editor in Chief
212-512-6242, fax 212-512-6111

Modern Plastics
McGraw-Hill Company, New York, NY 10020

Modern Plastics Encyclopedia
Modern Plastics, Raritan, NJ 08077

Motor Boating and Sailing
250 W. 55th St., New York, NY 10019-3203
212-649-3068, fax 212-489-9258

M&TS Journal
Newsletter of the International Machinery & Equipment Committee, American Society of Appraisers, Washington, DC 22070

National Equipment Digest
Vulcan Publications, Inc., P.O. Box 55886, Birmingham, AL 35255
800-877-9748

Natural Gas Week
1401 New York Ave., NW #500, Washington, DC 20005-2150
202-662-0700, fax 202-347-8089

New Equipment News
5080 Timberlane Blvd., 2nd Floor, #8 Mississauga, Ontario, LAQ 5C1 Canada
905-602-0814, 905-602-0818

The Official Railway Equipment Register
K-III Directory Corporation, 424 West 33rd Street, 11th Floor New York, NY 10001
800-221-5488

Oil & Gas Journal
1421 S. Sheridan, Tulsa, OK 74112
918-835-3161, fax 918-831-9497

Oil & Gas Investor
1900 Grant St. #400, Denver, CO 80203
303-832-1917, fax 303-837-8585

The Oil Daily
1401 New York Ave. NW #500, Washington, DC 20005-2150
202-662-0700, fax 202-783-8230

Parker's Evidence Code of California
Parker & Son Publications, Inc., Los Angeles, CA
Petroleum Engineer International
4545 Post Oak Place, #210, Houston, TX 77027-3105
713-993-9320, fax 713-840-8585

Plastics News
Crain, 1725 Merriam Road, Akron, OH 44313, Robert Grace, Editor
216-836-9180, fax 216-836-2322

Power Equipment Trade
1100 Superior Ave., Cleveland, OH 44114-2543
216-696-7000, fax 216-696-8208

A Primer of Oilwell Drilling
Petroleum Extension Service, The University of Texas at Austin, and International Association of Drilling Contractors

A Primer of Offshore Operations
Petroleum Extension Service, The University of Texas at Austin

A Primer of Oilwell Service and Workover
Petroleum Extension Service, The University of Texas at Austin and Association of Oilwell Servicing Contractors, Dallas, TX

A Primer of Pipeline Construction
Petroleum Extension Service, The University of Texas at Austin and Pipeline Contractors' Association, Dallas, TX

Principles of Engineering Economy
Grout Lresou and Leavenworth, John Wiley & Sons

Principles of Engineering Economy
The Ronald Press Co., New York, NY

Progressive Farmer
2100 Lakeshore Drive, Birmingham, AL 35209-6721
205-877-6000, fax 205-877-6450

Progressive Railroading
230 W. Monroe Street, #2210, Chicago, IL 60606
312-629-1200, fax 312-629-1304

Progressive Railroads
312-629-1200

Railroad Age
Simmons Boardman
312-427-2797

Railway Age
345 Hudson Street, New York, NY 10014-4590
212-620-7200, fax 212-633-1165

Railway Track & Structures
175 W. Jackson Blvd., Chicago, IL 60604
312-427-2729, fax 312-427-3014

Refrigeration and Air Conditioning
A. R. Trott, McGraw-Hill Book Company, New York, NY 10020

River Transport News
Criton Corporation, 9500 Crosby Road, Suite 200, Silver Spring, MD 20910
301-590-7180

Rubber and Plastic News
Crain, 1725 Merriam Road, Akron, OH 44313, Robert S. Simmons, V.P. Publications Director
216-836-9180, fax 216-836-2322

RUBBICANA Rubber Directory and Buyers Guide
Crain, 1725 Merriam Road, Akron, OH 44313
216-836-9180, fax 216-836-2322

Seatrade Weekly
Seatrade North America, Inc., 125 Village Boulevard, Suite 303, Princeton Forrestal Village, Princeton, NJ 08540
609-452-9410

Security Analysis - Principles and Techniques
Benjamin Graham, David L. Wood and Sidney Cotle, McGraw-Hill Book Company

Standard Industrial Classifications Manual
U.S. Office of Management and Budget

Successful Farming
1716 Locust St. Des Moines, IA 50309-3023
515-284-2853, fax 515-284-3127

Superintendent of Documents
U.S. Government Printing Office
Washington, DC 20402

Thomas Register
Thomas Publishing Company
Five Penn Plaza, New York, NY 10001
212-695-0500

Tire Technology International
U.K. & International Press, master House, Wescott, Dorking, Surry RH3NG, U.K.
44-0-306-743744, 44-0-642525

Trains
Kalmbach Publishing
800-533-6644

Urban Fiscal Stress
Lexington Books

U.S. Rail News
951 Pershing Drive, Silver Spring, MD 20910-4464
301-587-6300, fax 301-587-1081

Valuing A Company
George D. McCarthy and Robert E. Healy,
The Ronald Press Co., New York

Valuation of Property
Vols. 1, 2 and 4, James C. Bonbright, 1965, reprint by
Michie Company, Charlottesville, VA

The Vanderbilt Rubber Handbook
R.T. Vanderbilt Company, Inc., 30 Winfield Street,
Norwalk, CT 06855, Edited by Robert O. Babbit

Varied Monographs
American Society of Appraisers, Washington, DC

The Waterways Journal
Waterways Journal, Inc., 319 North 4th Street, Suite 650,
St. Louis, MO 63102, 304-241-4207
Who's Who in Electronics

Harris Publishing Company
Twinsburg, OH 44087

Workboat
Workboat, P.O. Box 908, Rockland, ME 04810-0908
304-241-4207

Workboat
3500 Highway 22, #205, Mandeville, LA 70470
504-626-0298, fax 504-624-4801

World Oil
P.O. Box 2608, Houston, TX 77252-2608
713-529-4301, fax 713-520-4433

World Plastics & Rubber
Cornhill Publications Ltd., 4-7 Nottingham Court, Shorts Gardens, London, WC2H9AY, John Murphy, Managing Editor
01-240-1515, fax 01-379-7371

Yachting
2 Park Avenue, New York, NY 10016-5601
212-779-5000, fax 212-725-1035

Index

A

B

C

D

E

F

G

H

I

J

K

L

M

N

O

P

Q

R

S

T

U

V

W

Y